THE DOPAMINE CODE

How to Rewire Your Brain for Happiness and Productivity

DR. SYDNEY CERUTO

Founder and CEO of MindLAB Neuroscience

Adams Media

New York   Amsterdam/Antwerp   London   Toronto
Sydney/Melbourne   New Delhi

Adams**media**

Adams Media
An Imprint of Simon & Schuster, LLC
100 Technology Center Drive
Stoughton, MA 02072

First Adams Media trade paperback edition June 2026

ADAMS MEDIA and colophon are registered trademarks of Simon & Schuster, LLC.

For information about special discounts for bulk purchases, please contact Simon & Schuster Special Sales at 1-866-506-1949 or business@simonandschuster.com.

The Simon & Schuster Speakers Bureau can bring authors to your live event. For more information or to book an event, contact the Simon & Schuster Speakers Bureau at 1-866-248-3049 or visit our website at www.simonspeakers.com.

Interior design by Colleen Cunningham
Interior images © 123RF/nunuu, hironicons

Manufactured in the United States of America

1 2026

Library of Congress Control Number: 2026935682

ISBN 978-1-5072-2611-7
ISBN 978-1-5072-2612-4 (ebook)

To Rachel,
for trusting me, for showing up in this work, and for believing in herself—
a quiet testament to what healing can look like, and a courage that
continues to inspire me every day.

ACKNOWLEDGMENTS

I'm deeply grateful to Brit Smith for believing in this vision and for championing me as the author of *The Dopamine Code*. Her insight, confidence, and editorial clarity from the very beginning made this book possible in the form it needed to take.

To Sarah Doughty, my editor, for her sharp eye, thoughtful guidance, and unwavering commitment to both the science and the reader. Her feedback sharpened every chapter and helped me translate complex neurobiology into something clear, practical, and human.

To the entire team at Simon & Schuster—editorial, production, design, marketing, and publicity—for their professionalism, care, and dedication to bringing this book into the world with integrity and impact.

To my colleagues, mentors, and neuroscience peers, whose work continues to deepen my understanding of dopamine, neuroplasticity, and the nervous system. Their research and clinical wisdom form the foundation of everything I teach.

To my son and my husband—for their patience, their love, and the quiet stability they provide. For the mornings and evenings, I was at the desk instead of fully present, and for the way they held space for this project without complaint. This book exists because they held the home.

Finally, to every client who has trusted me with their brain, their story, and their comeback. This work is for you.

Contents

PART 1

The Science of Dopamine—Understanding the Basics   17

CHAPTER 1

CHAPTER 2

CHAPTER 3

CHAPTER 4

PART 2

Under the Influence—
Navigating a World That Shapes Your Brain   75

CHAPTER 5

CHAPTER 6

CHAPTER 7

PART 3

The Dopamine Comeback—
Rediscovering Joy, Drive, and Lasting Satisfaction   117

CHAPTER 8

PART 4

Into the New Frontier—
Redefining Fulfillment, Enhancement, and Community   217

CHAPTER 15

CHAPTER 16

CONCLUSION

Preface

I used to believe I was simply restless, ambitious, and maybe a bit insatiable. But the truth was more complicated—and universal. I was, in every sense, a dopamine addict. My story wasn't defined by public unraveling. Instead, it played out in continuous, quiet craving: the thrilling rush of new love, the buzz of luxury and novelty, and the dissatisfaction that lingered no matter how much I achieved or acquired. While caught in this cycle, I was still showing up, however. I was raising my son, building my coaching practice, managing my household, and pursuing my academic interests. I was functioning, striving, and learning, while my brain chased the next high and crashed into the next low.

Getting deeper into the chase, I filled the void of a struggling marriage with shopping sprees, a boat, homes, luxury cars, six-figure vacations, and designer wardrobes. Each purchase brought a fleeting jolt of excitement, until a sense of emptiness returned, stronger than before. Still unsatisfied, I sabotaged comfort and security for a fresh high, limerence, and the intoxication of new romance and possibility. Yet as always, novelty faded. My patterns repeated: chase, acquire, and crash. I was trapped in a relentless race for the next thing, the next experience, the next high. Nothing ever lasted. Nothing ever satisfied me.

It was during a graduate semester on neuroscience, when I studied every brain chemical, that everything shifted. When the course arrived at the topic of dopamine, something inside me clicked into place. Suddenly, the endless race made sense. Those quick hits from constantly seeking novelty, the extreme lows that followed, and the pattern of living life in extremes all had an explanation. And where there was an explanation, there must also be a solution. I realized the answer wasn't in some outside novelty or big life change. The answer was within me. I could recalibrate. I could reset my brain. I could choose differently. I realized that my dopamine pathway—ancient, potent, and invisible—was steering my life. Only when I understood its script could I begin to rewrite my own.

What astonished me even more was discovering that many others—friends, clients, strangers—shared the same secret frustration, the sense that even life's brightest moments were shadowed by an inexplicable craving. That revelation changed everything. I began diving deeper into dopamine, connecting the dots between neurochemistry and behavioral patterns, and seeing more and more how it played into the lives of my clients. The more I learned, the clearer it became: Dopamine wasn't just about pleasure or addiction. It was about desire, anticipation, and the fundamental way humans move through the world. It was about why we crave more even when we have enough. My clients and countless others navigating similar struggles weren't broken. They were navigating an ancient biological system with modern tools—and in modern chaos. They needed not just awareness but a practical framework for building better habits that worked *with* their dopamine pathways rather than against them.

This book is that framework. It's what I've learned, tested, and refined through years of research and real-world application. And it's for anyone ready to stop being a passenger in their own story.

A note on the practice strategies in this book: Throughout *The Dopamine Code*, you'll find exercises designed to help you shift your habits and restore balance to your dopamine system. These are not prescriptive protocols that you need to complete in rigid order or sequential phases. Rather, they are a tool kit drawn from my twenty-five years of neuroscience-based coaching practice with high-performing professionals, executives in transition, and individuals seeking to reclaim emotional vitality after trauma. The journey to healthy dopamine pathways is not linear. Your nervous system has its own wisdom, and different brains respond to different interventions. I encourage you to approach these practices with genuine curiosity and honest self-awareness. Try them. Notice what shifts your system toward greater resilience, clarity, and dopamine balance. Pay attention to which strategies create relief or momentum. Build your personalized path from there, trusting your own experience as your guide. You may spend two weeks on one practice only to discover it's not the right fit for your dopamine system right now. That's normal (and valuable information). Try something else. Notice what happens.

My role through this book is to provide the science, framework, and options for harnessing dopamine. Your role is to get to know—and listen to—your own neurobiology. Turn the page to begin.

Introduction

Have you ever finished a big project only to find yourself feeling less satisfied than you expected? Are your best moments quickly eclipsed by the urge for the next upgrade, purchase, or notification? Or perhaps you constantly chase excitement or validation in your relationships, yet it never seems to be enough? If any of this sounds familiar, you're not alone.

Dopamine is the powerful chemical messenger in your brain that dictates how satisfied—or unsatisfied—you feel in daily life. But it isn't about pleasure alone—it's the engine behind every act, every avoidance, and every craving. Your ancestors relied on dopamine to get through famines, dangers, and uncertainty. Today, that same internal signal drives you to hit refresh, chase trends, and compare yourself to digital highlight reels. In relationships, too, the effects of dopamine run deep. Intimacy comes with those wonderful sparks of fulfillment, but it can be easy to get caught up in the search for the next spark, and the next—especially in a modern world where the possibility of someone "better" or "more exciting" is just a tap away.

The digital world and its abundance—apps, ads, online shopping, curated feeds—are designed to exploit those very circuits in your brain. With every ping, swipe, and scroll, your dopamine pathways are hijacked, making genuine contentment and even meaningful connections harder and harder to sustain. Fortunately, there is a way to rewire your own brain for real, lasting joy!

The Dopamine Code is your guide to the powers of dopamine, from uncovering how modern life plays on and influences these inner pathways, to establishing habits for deeper satisfaction and closer bonds. In Part 1, you will explore the science behind dopamine, including how dopamine kept early humans alive, and how it is fueled by anticipation. Then, in Part 2, you will dive into how modern technology can turn healthy pursuit into compulsive cycles in work, habits, romance, and creativity. You'll also look closer at how trauma can disrupt dopamine pathways and feed negative feelings that take away from pleasure.

In Part 3, you'll uncover the "Dopamine Menu" shaped in my own recovery and work with clients, and you'll learn how to create your own personal plan for well-being. You'll also practice hitting the reset button when your reward system is overwhelmed, and use resilience to overcome habits born from trauma. In Part 4, you'll look ahead at the future of dopamine, from how advances in neuroscience (the study of the brain and nervous system) will continue to impact this mini messenger, to how the Dopamine Revolution is shaping communities and fostering connection. Science isn't simply here to diagnose the problem—it will show you how to rewire your brain for sustainable happiness.

Throughout these parts, you will also find case studies. Amalgamations of the experiences from many clients I have worked with, these stories offer more insights into everything from the different impacts of dopamine to ways it can be better used for wellness. To make the most out of this book, there are also additional approaches to keep in mind as you read through each chapter, case study, and recommendation. Return to these whenever you need a refresher as you move through each part:

- **Read actively:** Each chapter following the Chapter 1 introductory content includes a "Pause and Practice" exercise—where science becomes concrete action in your day. Try each exercise before moving onto the next chapter. (Keep in mind that while some guidelines for time frame, order of steps, etc. are provided, how and when you use this practical advice is up to you and your needs.)

- **Experiment bravely:** Test new routines. Rewiring your brain's pathways takes time, but every tweak strengthens habits that support lasting joy.
- **Share the journey:** These strategies are springboards for conversations—bring a friend, partner, or colleague along. Change is accelerated in community.
- **Design, don't drift:** Shape your environment to support your goals. Change your phone settings, your workspace, your family rituals—transform your landscape for joy.
- **Honor progress and imperfection:** Growth is nonlinear. Setbacks teach, routines evolve, and satisfaction deepens as you adapt.

If you've felt hijacked by craving, burnout, or distraction—if happiness, satisfaction, and fulfillment seem "almost" but rarely "really" yours—you've come to the right place. The pages ahead hold the blueprint—not for perfection, but for lasting, authentic change grounded in your brain's true strengths. Your brain hasn't failed you; it's waiting for you to break the dopamine code. And I promise: The solution is within reach. Let's dig in.

PART 1

The Science of Dopamine

Understanding the Basics

Dopamine is a chemical messenger (or neurotransmitter) produced in the brain that plays a crucial role in regulating movement, motivation, reward, learning, and emotional responses. While most people think of dopamine as the "pleasure molecule," that's only a fraction of the story. In reality, dopamine is a central force for healthy brain function, shaping how you think, feel, and act every day. It's the signal that pushes you to pursue goals, form new habits, and overcome obstacles, even when the odds are steep.

In this part, you'll learn more about dopamine and how it works. Across the next four chapters, you'll debunk common myths that have shaped public conversations about dopamine, dive deep into the science behind how habits form and break, explore why you chase some rewards endlessly while others quickly lose their spark, and learn practical strategies rooted in neuroscience for rewiring routines and boosting self-control.

You'll see how dopamine shapes your cravings and choices—from your first cup of coffee to your most ambitious life goals. You'll uncover why some habits stick while others are lost to distraction, and you'll learn the principles for ultimately taking charge of your own motivational system. By the end of this part, the science won't just be fascinating—it'll be helpful in the next step of navigating change, developing resilience, and transforming the way you think and act.

You'll also see how technology and culture ingeniously shape what your brain craves, impacting everything from daily habits to what you value most. And through case studies, you'll recognize the tug-of-war between wanting and true satisfaction, the subtle ways your environment teaches you to chase, and the exhaustion that comes from pursuing ever-bigger "hits" until pleasure itself seems out of reach. As you read through this part, you'll get a foundation for genuine self-understanding that's rooted in neuroscience and ready to be put to work in the following parts.

CHAPTER 1

What Is Dopamine?

The Molecule of Motivation

Dopamine is the central force behind anticipation, motivation, and goal-directed behavior—the energy that moves you forward, prompts curiosity, and fuels ambition. Every moment you strive for something, whether it's a career milestone, a deeper relationship, or simply the next possibility around the corner, dopamine is at work shaping your focus and your drive.

This chapter reveals how dopamine truly influences your daily life. You'll dig deeper into how dopamine isn't merely about pleasure, but instead underpins the urge to seek, imagine, and persist—even in the face of obstacles. You'll discover how the dopamine system operates in the brain, orchestrating motivation through dynamic neural circuits. And by looking through the lens of human evolution, you'll understand why fulfillment always seems just out of reach and why your brain was designed to prefer anticipation over satisfaction.

You'll also see how dopamine's influence extends to relationships, resilience, and identity. Whether you thrive on novelty or prefer steady routines, understanding your unique dopamine profile will empower you. You will also meet Johan, whose story illustrates that even the highest achievers can struggle when their reward system's baseline shifts. By the end of this chapter, you will have the tools to move beyond fleeting urges and start cultivating the same meaningful, lasting joy that Johan eventually rediscovered.

Beyond the "Pleasure Chemical": The True Nature of Dopamine

Popular culture loves to label dopamine as the "pleasure molecule," yet science reveals a richer, more nuanced reality. Dopamine's true role is found not simply in the thrill of reward, but in the dance of anticipation, striving, and imagining what could be. The surge of motivation that energizes you at the start of a new day, the focus that allows you to solve complex problems, and the curiosity that fuels ambition are all orchestrated by dopamine.

Dopamine doesn't give contentment; it delivers possibility. It sets the stage for new challenges, frames future goals as attainable, and propels you to ask "What if?" about new opportunities, relationships, or aspirations. The anticipation of a meaningful conversation, the excitement for a fresh project, and even just the allure of a delicious meal are all sparked by dopamine's drive for future reward.

Critically, dopamine rewards the process, not just the finished result. It's what sharpens focus and fills ordinary moments with potential. Without healthy dopamine signaling, life can seem colorless and static, lacking drive and excitement. Your brain, not designed for endless satisfaction, continually seeks out what's next. Even in your busiest or most restful moments, dopamine is at work. After completing a big project or celebrating a milestone, you may notice that sense of achievement quickly fades, replaced by a desire for what comes next. This cycle isn't a personal deficiency—it's the brain's adaptive drive, always seeking possibility over static contentment. Understanding this will be your first step to seeking out challenging experiences intentionally and appreciating the anticipation itself as a vital part of daily satisfaction (more on this in Part 3).

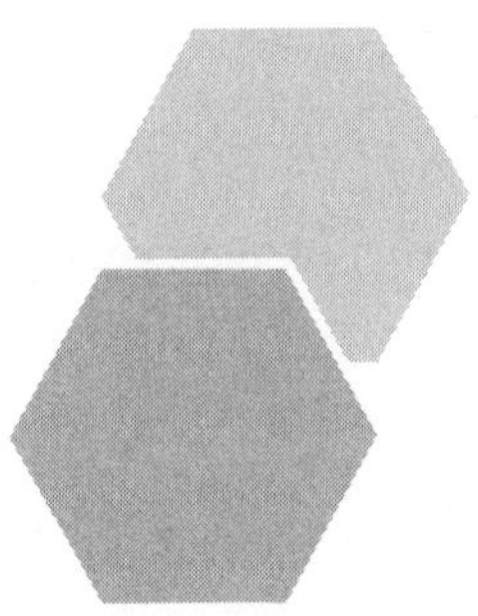

Dopamine doesn't give contentment; it delivers possibility.

It sets the stage for new challenges, frames future goals as attainable, and propels you to ask "What if?" about new opportunities, relationships, or aspirations.

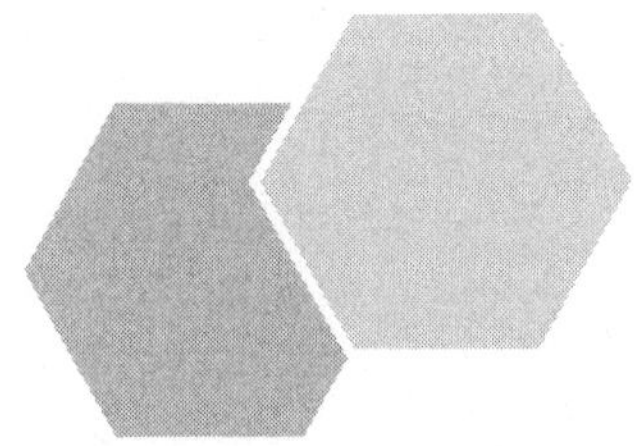

Dopamine's Circuit: An Orchestra of Anticipation

Now that you've looked closer at what dopamine is, the next part of understanding this internal messenger is exploring how exactly it works. Dopamine does not act alone; it moves through a complex series of **neural circuits**, coordinating various regions of the brain to turn your desires into action. Imagine your brain like an orchestra, with dopamine serving as the conductor, directing who plays and when, making sure every section comes together to produce something greater than the sum of its parts.

A CLOSER LOOK
Neural Circuits

You can think of neural circuits as the brain's communication networks. They are groups of cells in your nervous system that work together to process specific information and control your behavior. Each circuit acts like a feedback loop: One cluster of cells sends signals to another, which then sends responses back, shaping how you think, feel, and act.

The process for this internal orchestra often begins deep within the **midbrain,** in a region called the **ventral tegmental area** (**VTA**). Here, dopamine is first released in response to signals of potential or possibility, like the start of a goal, or the first inkling of curiosity. These dopamine signals are then sent to the **nucleus accumbens**, which is sometimes referred to as the brain's "reward center." This area is responsible for reinforcing positive experiences: It shapes what you find pleasurable and helps you remember and seek out those rewards again.

🔍 A CLOSER LOOK
The Midbrain, Ventral Tegmental Area, and Nucleus Accumbens

Midbrain: The control center deep inside your brain that bridges your basic instincts with your higher thinking. It's where motivation begins, detecting when something matters to you and triggering the dopamine release that gets you moving toward what you want.

Ventral tegmental area: A small region in the midbrain densely populated with dopamine-producing neurons. The VTA is the primary source of dopamine release in the brain and serves as the starting point for motivation and reward-seeking behavior.

Nucleus accumbens: Part of the brain's reward system, this region essentially learns what feels good and drives you to pursue those experiences again and again, making it central to habit formation and goal-directed behavior.

But the dopamine journey does not stop there. The **prefrontal cortex**—the area of your brain that governs reasoning, planning, and focus—receives these signals and weighs whether a reward is worth pursuing now, or whether it pays to wait for something greater in the future. This interplay allows you not only to want, but to strategize—to chase a long-term goal instead of settling for short-term gratification.

🔍 A CLOSER LOOK
Prefrontal Cortex

The prefrontal cortex is also known as the executive hub of your brain and is located just behind your forehead. In addition to reasoning, planning, and focus, it also handles impulse control.

When this orchestra is in harmony, you experience motivation, productive focus, and satisfaction as you move through your day. But if this dopamine system becomes imbalanced—overactive or underactive—you might notice certain effects in subtle or dramatic ways. You may feel a loss of drive, difficulty focusing, impulsiveness, or even a low mood that is tough to shake. Too little dopamine and ambition dwindles; too much, and you may become restless, easily distracted, or find yourself chasing novelty for its own sake.

The Evolutionary Story: Why Your Brain Runs on Anticipation

To truly grasp dopamine's influence, it helps to step back and see your brain through the lens of human evolution. Your earliest ancestors confronted daily uncertainty—survival was never a guarantee. Each day brought dangers and opportunities: tracking food, seeking shelter, or building relationships that could make the difference between success and failure. In that world, sitting still or feeling satisfied for long could put your future at risk.

Dopamine evolved to help humans thrive in this constant state of "not quite enough." It became the biological engine of anticipation—fueling the curiosity to scout new terrain, the perseverance to try again after setbacks, and the drive to form bonds that boosted collective survival. What the brain truly cherished was not comfort, but the sensation of moving forward, of working toward a goal just out of reach.

The most powerful dopamine surges came *before* rewards were won. Planning a hunt, imagining the taste of a meal, or plotting ways to win another's trust all sparked motivation that kept your ancestors trying even when the outcome was uncertain. The satisfaction at the finish line was fleeting, but the energy to keep striving was critical: That's the legacy dopamine left in every modern brain.

Fast-forward to today's world, and you'll see these same circuits at work. You are still wired to crave challenge, exploration, and the hope of a reward that lies just ahead. It's why boredom creeps in so quickly, why completing a significant goal is soon followed by restlessness, and why some people are drawn to novelty or risk as soon as comfort sets in. Your brain did not evolve to bask for long in satisfaction; anticipation is your default mode, not a design flaw.

This evolutionary inheritance explains so much about the human experience: getting bored even when you "have it all," or feeling most alive in the pursuit rather than the win. When you recognize that the search for what's next is part of your design and not a personal failing, you can begin to work with it and harness anticipation to motivate growth and resilience. In the next section, you'll turn a critical eye to how abundance, constant stimulation, and easy rewards are affected by this ancient drive, and what this means for joy and productivity in your everyday life.

Modern Abundance: Navigating Desire in a World of Plenty

Today's world is one of almost continuous stimulation and endless opportunity. Instead of braving hunger, danger, and environmental threats, many people now face a barrage of notifications, unlimited choices, and instant gratification at every turn. In theory, this abundance ought to make them happier and more satisfied than any earlier generation. Yet, for many people, it delivers the opposite: a persistent sense of emptiness or restlessness just below the surface.

This isn't a moral failing or a lack of willpower. It's a natural side effect of a brain hardwired for anticipation, now immersed in an environment overflowing with chances to want, chase, and consume. Every ping, trending headline, or potential reward triggers a spark of dopamine, priming you to seek out more. But when these small jolts arrive too frequently and with little real effort, your dopamine system doesn't have time to reset. Over time, this constant exposure blunts the system, and what once excited you—a new message, a small prize, an accomplishment—now demands ever greater stimulation just to feel a faint response.

You will dig deeper into the impacts of the modern world on your dopamine pathways in Part 2, but it's helpful to have an overview when looking at the evolution of the brain and dopamine. How your ancestors were shaped by anticipation is important insight into where your own wants and needs come from.

Dopamine: A Tapestry of Connection, Grit, and Identity

Of course, dopamine's reach extends far beyond your personal wants and urges—it's fundamentally intertwined with your relationships with others, your capacity for resilience, and how you construct your very sense of self. The anticipation of social connection, acceptance, or belonging can activate your brain's reward system just as powerfully as the pursuit of personal goals. This is no accident of culture; the drive to be seen, valued, and included is a biological imperative cemented by that evolutionary "inheritance" you learned about earlier in this chapter. As mentioned previously, social bonds increased chances of survival for your ancestors; today, these bonds remain essential for emotional and mental well-being.

Think of the thrill of new romance, the comfort of laughter with close friends, or the pride of being recognized in a group. All these experiences are orchestrated by dopamine. Your brain lights up not only for tangible achievements, but also for the social rewards that come from connection and shared experience. In fact, much of your day-to-day motivation is driven by the anticipation of social belonging—whether you're preparing for a family gathering, collaborating at work, or hoping for acknowledgment from someone you respect.

Dopamine also shapes the stories you tell yourself about who you are and what you are capable of. The way you interpret successes and setbacks—your narrative of grit and reinvention—is deeply influenced by how your dopamine system encodes expectation and surprise. For instance, when a plan unravels or a hope is dashed, a dip in dopamine can spark uncomfortable feelings of restlessness or frustration. Yet, this drop is a biological nudge, urging you to reevaluate, adapt, and try again. In neuroscience, this process is known as the **reward prediction error**, and it transforms disappointment into an opportunity for learning and growth.

🔍 A CLOSER LOOK
Reward Prediction Error

> Reward prediction error is the gap between what you expected to happen and what actually happened.

It's in these moments—where expectation meets reality and the outcome is uncertain—that you may find your most significant potential for self-understanding and reinvention. The ability to persist after setbacks, to recover your optimism, or to try a different approach is fundamentally a dopamine story. Over time, how you respond to these dopamine-driven signals shapes your sense of courage and ability to overcome challenges.

Everyone's relationship with dopamine is unique. Genetics, past experiences, and personal tendencies all influence how you are wired: Some people thrive on novelty and risk, while others find deep satisfaction in routine, mastery, and stability. Recognizing your own "dopamine landscape" can transform self-judgment into self-compassion, empowering you to align your efforts with what truly motivates you.

CASE STUDY
Johan—The Diminishing Return of Success

Johan held a senior position at his company. His career path had brought him everything he'd set out to earn: respect, international accolades, and a comfortable lifestyle. For years, each new challenge at work fueled his sense of purpose. Every professional achievement felt meaningful.

Yet, as his successes accumulated, something shifted. Johan began to notice that the thrill of achievement didn't last. Even monumental victories, like winning industry awards, left him oddly deflated. Eventually, he found himself on autopilot at work and disengaged at home. Family moments, once joyful, now felt muted; weekends blended together in a haze of distraction. "I'm not unhappy," he confessed. "But I don't feel much of anything. Each good thing just fades."

In our sessions, we examined Johan's patterns and routines. The common thread wasn't a lack of gratitude or ambition—it was overstimulation. Years of constant reward and novelty had conditioned his dopamine system to expect ever-higher peaks just to register excitement. Achievements that would have thrilled him years earlier now barely made an impression. His story wasn't about burnout in the traditional sense; it was about the steady, invisible lowering of his reward system's baseline—the neuroscience of "dopamine downregulation."

To interrupt this cycle, we mapped out small but meaningful resets. Johan introduced "novelty fasts," deliberately abstaining from luxury and convenience for several days a week. He walked or biked around his home city, sought out new (yet simple) experiences, and disconnected from screens after work. Professionally, he shifted focus from how much work he was producing to creative solution-finding—mentoring younger staff, partnering with community groups, and designing experiences for the company's clientele.

At home, he invested in learning and family connection: helping his daughter cook new recipes, trying improv nights with his spouse, and opening his social circle to friends beyond the workplace. These activities, sometimes awkward or unfamiliar at first, gradually reawakened his anticipation and sense of accomplishment.

Johan's journey underscores a universal truth: Lasting motivation and joy come not from piling up rewards, but from regularly seeking challenge, meaning, and connection. By intentionally disrupting old patterns, he restored his capacity for anticipation—and, with it, his appreciation for both work and life's everyday moments.

Shaping Your Own Dopamine Menu: Science to Practice

Understanding dopamine is the first step; changing your relationship with it requires intentional, practical shifts. In Part 3, you will use your knowledge of dopamine from Part 1, as well as the obstacles you explore in Part 2, to design a tool kit of habits, experiences, and mindsets that keep your motivation healthy rather than depleted. This "Dopamine

Menu" will be specific to the personal needs and wants you uncover, as well as the larger social motivations of your community. As you move on to the next chapters, consider how everything you discover is building toward this personalized Dopamine Menu.

CHAPTER SUMMARY

As you've started to discover, dopamine is a complex molecule responsible for many things. This is why each chapter in this book ends with a list of key takeaways. The following are the main points to remember from this chapter as you continue your journey to decode dopamine:

- Dopamine is the biological basis of motivation, anticipation, and sustained satisfaction, not just brief pleasure.
- The paradox between wanting and liking is not a design flaw but an evolutionary adaptation: The survival of your ancestors required relentless pursuit, not prolonged contentment.
- Modern life's constant stimulation can dull the dopamine reward system, but deliberate "resets" and challenge can restore balance.
- True agency and fulfillment emerge by understanding your dopamine landscape and designing your habits accordingly.

CHAPTER 2

The Dopamine Paradox

Wanting versus Liking

Why do people chase rewards that so often leave them unsatisfied? The paradox at the heart of dopamine's role in the brain is this: Wanting and liking are not the same. Neuroscience reveals a striking divergence between the two: Wanting is an intense, future-focused anticipation, while liking is the actual experience of pleasure and is often brief and muted. You live in a world overflowing with opportunities to seek, strive, and achieve; yet, the actual enjoyment of these rewards can be elusive or fleeting.

This chapter unpacks why humans drive themselves to chase moments, accolades, and possessions that may never truly deliver fulfillment, and why disappointment and restlessness follow even significant successes. You'll learn about the underlying mechanisms that separate desire from contentment; explore how dopamine wires the brain for anticipation, risk-taking, and persistent pursuit; and uncover how dopamine is far less involved in producing lasting satisfaction. You'll also delve into the science of reward prediction, the thrill of uncertainty, and how the "maybe" factor—unpredictable outcomes—supercharges motivation but rarely guarantees happiness. In addition, you'll discover how biology, hormones, and sociocultural influences can have different impacts on everyone, making this journey to decoding dopamine unique to you.

Through the lens of lived experience, you'll meet Viviana—a high-achieving project manager whose accomplishments failed to immunize her against burnout and emotional numbness. Her path mirrors the struggles of millions stuck on the treadmill of achievement, illustrating how persistent craving can overshadow even life's brightest moments. In this chapter, you'll acquire practical strategies for distinguishing wanting from liking, rebalancing your dopamine system, and making choices that foster authentic fulfillment. By the end, you'll possess a clear understanding of how to turn the paradox of dopamine into a road map for deeper, more satisfying rewards in every area of your life.

The Dopamine Paradox: Why Wanting Isn't Liking

Dopamine's reputation as the "motivator" is rooted in how it wires people to pursue, imagine, and strive. Yet, while it supercharges anticipation, it doesn't guarantee lasting pleasure. There is a split between *wanting* something and *liking* it once it is acquired: The rush that comes before a goal rarely matches the contentment felt after achieving it. Have you ever noticed the delight of a new purchase, promotion, or personal win fade quickly? That's dopamine at work—gearing up for the subsequent pursuit of something else even as the current reward barely registers. At the neurobiological level, dopamine floods your brain circuits as you *anticipate* ("want") rewards, especially in regions like the nucleus accumbens and **ventral striatum** (the reward-seeking hub of your brain that lights up when you're chasing something you desire). The actual *enjoyment* phase—the "liking" of a reward once you have it—is supported by other chemicals: **endogenous opioids**, **endocannabinoids**, and **serotonin**. Their effect is more subtle than the drive of dopamine. The brain's architecture means you're energized not by what you have, but by what lies ahead.

The reward prediction error highlights yet another quirk of your brain. Touched on in Chapter 1, this error is worth a deeper look here, as it helps shed light on the differences between wanting and liking. When an expected reward is delivered, dopamine neurons briefly quiet; if the reward surpasses expectation, activity jumps. But if the outcome disappoints, dopamine drops quickly, fueling frustration and quickly rerouting the pursuit elsewhere. This dynamic can be seen in phenomena like buyer's remorse and post-achievement burnout, and in why so many people are addicted to "the chase." Consider how these principles play out in relationships, creativity, and goal setting. When wanting dominates, partners may chase novelty, overlook stability, and undervalue small moments of connection. In creative work, the thrill of starting something new is high, but completing that project can feel anticlimactic.

A CLOSER LOOK

Endogenous Opioids, Endocannabinoids, and Serotonin

While dopamine drives pursuit, these three neurochemical systems create the actual sensation of pleasure and contentment. Together, they produce the warm glow of genuine enjoyment, though their effects are subtler and shorter-lived than dopamine's forceful drive.

Endogenous opioids: Your brain's natural painkillers that produce a sense of euphoria and deep satisfaction.

Endocannabinoids: Generate calm and ease.

Serotonin: Fosters mood stability and the quiet sense that all is well.

Understanding the distinction between wanting and liking is transformative. It explains why *chasing* more rarely leads to *feeling* more—and why sustainable well-being requires experiences, mindsets, and habits that cultivate genuine liking alongside healthy pursuit. People are often unwittingly trapped on a treadmill of acquiring, striving, and consuming—always seeking the next dopamine hit while feeling mysteriously dissatisfied.

Tracing the Evolutionary Roots of the Dopamine Paradox

As you explored in Chapter 1, evolution has played a key role in the development of the dopamine system. The split between wanting and liking is a part of these evolutionary roots. Early humans didn't have supermarkets, climate control, or an endless menu of entertainment at their fingertips. Survival required a bias toward relentless pursuit. If your ancestors had been entirely content after a single good hunt, they might not have gone out again to secure their next meal. Dopamine's "wanting" system evolved to prime action before necessity struck, keeping humans moving toward resources, mates, allies, and safer territory.

The "liking" system, by contrast, was never meant to last long. A quick surge of pleasure rewarded success, but lingering in that state would have invited vulnerability to predators or missed future opportunities. In neurobiological terms, the brevity of liking ensured that the wanting system was quickly re-engaged after a win. Anthropological evidence supports this: Hunter-gatherer groups often celebrated immediately after a big hunt, then went straight into planning the next expedition, sometimes within hours.

These ancient patterns now play out in a very different environment. The same circuitry that kept the species alive propels people to refresh their email inbox, chase promotions, and scroll endlessly through social media. The brain doesn't distinguish between tracking prey across the savanna and acquiring "likes" on an *Instagram* post; both register as potential rewards. By understanding this evolutionary mismatch, you can begin to break the cycle. When you feel the restless urge for "what's next" just moments after a win, it's not a personal weakness; it's a Stone Age survival tool working in a twenty-first-century world.

Separating Fact from Fiction in the Dopamine Paradox

In popular discourse, dopamine is often misunderstood or oversimplified. It's blamed for addiction or described as the magic ingredient behind motivation. The reality is much more nuanced. Dopamine does not directly create pleasure or happiness; instead, it orchestrates wanting, pursuit, and anticipation—fueling drive, curiosity, and risk-taking. High dopamine levels don't always translate to high satisfaction; in fact, they can create persistent craving and restlessness, leading to cycles of over-work or compulsive behaviors. Meanwhile, low dopamine isn't synony-mous with apathy or laziness alone; it may signal burnout, chronic stress, or even just a temporary **downregulation** after periods of high effort.

True contentment involves a balance of dopamine with serotonin (mood regulation), **endorphins** (pain relief and euphoria), and **oxyto-cin** (connection and bonding). Tangible rewards—those that foster both wanting *and* liking—depend on this balance. Recognizing these truths can help you understand your motivation without stigma or self-blame.

A CLOSER LOOK
Downregulation

Downregulation is the process by which your brain becomes less responsive to a stimulus—in this case, dopamine. When dopamine floods your system repeatedly, your brain cells adapt by producing fewer dopamine receptors or reducing their sensitivity. This means you need more of the same stimulus to feel the same effect. Downregulation is why the thrill of a new purchase fades, why social media scrolling requires more time to satisfy, and why chasing the next win feels like running on a treadmill that never stops.

The Dopamine Paradox and ADHD

For those with attention-deficit/hyperactivity disorder (ADHD), the split between wanting and liking is even more pronounced. The ADHD brain has higher concentrations of **dopamine transporters**, which are proteins that remove dopamine from active circulation. In the ADHD brain, these transporters clear dopamine too quickly from the **synapse**, the tiny gap where one brain cell communicates with another.

Individuals with ADHD often discover that what others achieve through steady motivation, they must orchestrate through deliberately structured environments, accountability, novelty injection, and exter-nal reward systems.

The ADHD brain doesn't simply crave rewards differently; it experiences a particular version of the dopamine paradox that creates both an extraordinary gift *and* persistent frustration. People with ADHD often exhibit hyperfocus: an intense, almost magnetic ability to lock on to stimulating activities, driven by surges of dopamine anticipation. In these moments, they feel alive, purposeful, and utterly absorbed. Yet this intense desire comes with a twist: Once the novelty fades or the task is completed, satisfaction rarely lingers. Instead, the brain quickly turns to chasing the next stimulating experience.

What compounds this struggle is that for those with ADHD, many everyday tasks like routines, administrative work, and mundane responsibilities don't generate enough dopamine production to activate that motivational engine. The result is a brain wired to pursue intense wants yet starved for the reliable liking that sustains meaningful progress across work, relationships, and personal growth.

A CLOSER LOOK
Dopamine Transporters and the Synapse

Dopamine transporters: Specialized proteins embedded in cell membranes that act like cleanup crews, removing dopamine molecules from the space between neurons. In non-ADHD brains, this recycling process maintains steady dopamine signaling. In ADHD brains, an overabundance of these transporters aggressively clear dopamine, leaving insufficient dopamine lingering in the synaptic space.

The synapse: The junction between two brain cells where chemical communication occurs. When one neuron fires, it releases neurotransmitters (like dopamine) across this gap to activate receptors on the receiving cell. The synapse is where the actual conversation between neurons happens, and its efficiency determines how strong and clear that communication is. For those with ADHD, the synapse does not transmit dopamine's message effectively, compelling the brain to seek more intense or novel stimulation in order to get a strong response.

Just understanding this nuance can dissolve much of the shame surrounding ADHD. It's not a matter of laziness or lack of discipline; it's a fundamental difference in how the reward prediction system operates. Once it is recognized that ADHD represents an extreme expression of the dopamine paradox (an exceptional capacity for craving and simultaneous difficulty sustaining satisfaction), those with ADHD can start designing strategies that work with this wiring rather than against it. The challenge can be turned into a framework for realistically managing the brain's impulses and embracing genuine achievement.

Desire and Drive: Sex Differences in Dopamine

Groundbreaking research in **neuroendocrinology**, the study of how hormones influence the nervous system and brain function, reveals why men and women often pursue, experience, and sustain rewards differently.

○ A CLOSER LOOK
Neuroendocrinology

Neuroendocrinology is the field that examines how hormones and the nervous system interact to shape behavior, emotion, and cognition. This field helps explain why motivation, pleasure, and satisfaction can vary across individuals, life stages, menstrual cycles, and between sexes.

Testosterone and estrogen modify how dopamine circuits fire, adapt, and recover. In men, testosterone increases **dopamine receptors**—proteins that receive dopamine signals and activate motivation—in the brain's reward areas. This effect sharpens initiative, risk tolerance, and the drive to compete with others. This effect is supercharged during puberty but continues to modulate drive and resilience throughout adulthood.

⚲ A CLOSER LOOK
Dopamine Receptors

Dopamine receptors are designed specifically to receive dopamine molecules. When dopamine is released, it seeks out and binds to these receptors, activating them like a key turning in a lock. The more receptors present on a cell, the more dopamine signals can be received and amplified.

Women's reward processing, on the other hand, displays cyclical changes. Estrogen, highest around the ovulation stage of the menstrual cycle, boosts dopamine sensitivity, increasing curiosity and responsiveness to both social and achievement-oriented rewards. Progesterone (highest after ovulation) and other modulators such as **allopregnanolone** (enhances calming GABA activity), **GABA** (functions as the brain's primary calming neurotransmitter), and **oxytocin** (rises during social bonding and nurturing behaviors), add layers of nuance, sometimes tempering the drive for novelty with more attention to emotional and social context. As a result, studies find that women may pursue achievement not only for its own sake, but often for its impact on relationships, stability, or collective well-being.

Societal influences further shape these sex-specific patterns. In certain cultures, boys are often praised for risk-taking, assertiveness, and independence, which in turn reinforces dopamine-driven behaviors. Meanwhile, girls may receive messages—explicit and implicit—encouraging collaboration, empathy, and balanced ambition. These messages reinforce connection-based and oxytocin-driven behaviors alongside more tempered dopamine activation.

Hormonal changes across the lifespan also play a role. During perimenopause and menopause, the drop in estrogen can affect motivation and energy. In men, declining testosterone in later life may impact drive, while self-confidence often shifts toward relational and mentoring satisfaction.

Despite these trends, however, each person is different: Many women are high-achievers and risk-takers, and many men find deep fulfillment in connection and care. Individual experience always trumps stereotypes. Those who self-reflect, challenge inherited narratives, and cultivate a values-based approach to achievement and pleasure find far greater satisfaction. Healthy reward systems emerge in environments that support diversity, celebrating both competitive success and the joys of community, collaboration, and care.

A CLOSER LOOK
Allopregnanolone, GABA, and Oxytocin

While dopamine drives pursuit and desire, other neurochemicals provide balance:

Allopregnanolone: An element of the metabolism of progesterone that acts as a powerful calming agent in the brain, reducing anxiety and promoting relaxation.

GABA (gamma-aminobutyric acid): The brain's primary inhibitory neurotransmitter that puts the brakes on neural activity and creates a sense of calm and ease.

Oxytocin: Often called the "bonding hormone," this is a neurochemical that promotes social connection, trust, and nurturing behavior. Unlike dopamine, which drives pursuit and desire, oxytocin creates feelings of safety, belonging, and satisfaction in relationships.

The interplay of hormones, personality, and social learning shapes a unique **dopamine profile** for each person, regardless of sex. And this profile can be a powerful tool for personal growth. By tracking your own personal motivation cycles, practicing self-acceptance when drive fluctuates, and nurturing relationships that reinforce both achievement and connection, you can move from comparison and frustration toward empowered self-design.

A CLOSER LOOK
Dopamine Profile

Your dopamine profile is your unique neurobiological pattern of dopamine sensitivity, reactivity, and regulation. It encompasses how readily dopamine is released in response to rewards, how many dopamine receptors you have, how quickly your body recycles dopamine, and how your dopamine system responds to stimulation and stress. Your dopamine profile is shaped by genetics, hormones, life experiences, and learned behaviors. Understanding your personal dopamine profile reveals your natural tendencies toward risk-taking, motivation, novelty-seeking, reward sensitivity, and how easily you experience satisfaction. This self-knowledge empowers you to work with your biology rather than against it, designing environments, goals, and practices that align with how your brain is wired.

You will uncover your unique profile as you move through this book and learn how to use it to create a personalized Dopamine Menu in Part 3.

Culture and Reward: The Social Brain

Humans are social beings, wired to crave connection, status, feedback, and inclusion. As you learned in Chapter 1, dopamine isn't specific only to personal urges and goals but also extends to relationships with others. Digging even deeper, dopamine is tuned by culture and community, and these factors play a big role in shaping what a person experiences as "rewarding." Beyond more specific ideals based on sex, every society as a whole values certain goals—wealth, success, relationships, virtue—and the dopamine system is conditioned over time to pursue those rewards with urgency.

In Western cultures, individual accomplishment is a dominant narrative: career promotions, material gain, athletic wins. Dopamine surges in response to personal milestones, but so does craving for peer recognition, social proof, and public validation. In collectivist societies, such as Japan, South Korea, and many Latin American communities, reward is often tied to group achievement, reputation, and legacy. The anticipation of shared success or community harmony can even outstrip private ambition.

Media, especially digital platforms, exploit this "social dopamine." The notification of getting a "like" on social media, the thrill of your content going viral, or the anticipation of an online message all activate reward pathways, leading people to compulsively check and persistently engage with media. Further, social comparison—fueled by curated highlights and influencer culture—triggers envy and restlessness, potentially undermining satisfaction even when a person is successful.

At work, the social signals for dopamine are equally powerful. Recognition, inclusion in key projects, mentorship, and networking can be more motivating than promotions or bonuses. Exclusion, conversely, registers as reward deficiency, fueling craving for connection.

Strategies for reclaiming healthy dopamine levels in a cultural context (which you will explore more in Part 3) include resisting the pressure to pursue what "everyone" wants, being mindful of the digital media you consume, and dedicating time to authentic relationships. Family rituals, collective service projects, and group celebrations can provide more lasting satisfaction. Evidence from **positive psychology** (the scientific study of well-being and the conditions that enable people and communities to thrive) demonstrates that communities marked by connection and recognition have healthier dopamine balance and lower rates of burnout.

The **neurobiology** of cultural reward isn't destiny: You can consciously choose which signals to amplify and which to reject.

By regular reflection—journaling on which rewards matter most, pausing to savor real connections—you can cultivate social environments that nourish genuine liking, not just habitual wanting. You will continue to reflect and learn more about building helpful connections as you move through this book.

🔍 A CLOSER LOOK
Neurobiology

Neurobiology is the science of how the brain and nervous system work. It examines the structure of different parts of the brain regions, how neurons communicate, and how brain chemistry influences behavior, emotion, and decision-making. When I talk about the "neurobiology" of reward or motivation, I'm looking at the actual physical mechanisms in your brain that make you feel driven, satisfied, or craving more.

Reward Uncertainty: The Thrill of Maybe

Neuroscientists have shown that the largest dopamine spikes come from unpredictable rewards—not the ones expected, but the ones that *might* happen. This principle, called **variable reinforcement**, explains why gambling, dating apps, and social media are so compelling. The possibility—rather than the guarantee—of winning a hand, getting a private message, or of other positive outcomes drives people to persistently seek that variable reinforcement. And it can create powerful habits.

🔍 A CLOSER LOOK
Variable Reinforcement

Variable reinforcement is a pattern of reward where the outcome is unpredictable. Instead of receiving a reward every time you perform an action, you get rewarded randomly or on an unpredictable schedule. This unpredictability is powerfully motivating to the brain because dopamine surges highest when you're uncertain whether a reward will come. Slot machines, phone notifications, and more exploit variable reinforcement.

In studies with animals (such as research by Fiorillo et al., January 2003), the subjects continued to press a lever when it was uncertain whether they would receive a reward for pressing it. Meanwhile, they stopped pressing the lever when it was a guarantee that they would receive a reward: Certainty led quickly to boredom. Human brains work similarly: Uncertain praise at work, unpredictable attention from loved ones, fluctuating social media feeds—all produce more prolonged, more intense dopamine activation than predictable rewards. The anticipation and hope are magnetic, but the pleasure on arrival is mild and brief.

Uncertainty can be a good thing. Namely, it propels creativity and innovation: Artists, entrepreneurs, and explorers are often driven by "What if?" The search for breakthrough, discovery, or unique experience is advantageous, even as arrival can feel anticlimactic. The healthiest dopamine system is one that can tolerate uncertainty, take risks, and persist in the absence of immediate results, but also pause and savor when the outcome is achieved. Too much uncertainty, however, strains well-being. Chronic ambiguity in work roles, relationships, or personal identity creates anxiety and distraction and undermines real contentment. The lesson is balance: Seek novelty and unpredictability with intention, but anchor your life in practices that provide regular satisfaction, stability, and peace.

Practical Strategies: Mapping Your Dopamine Patterns

Pause and practice: Before moving forward, take a few minutes here to explore how the concepts in this chapter show up in your own life. This simple reflection will help you spot the gap between *wanting* and *liking*—and give you valuable insight into where your true satisfaction comes from. You can simply reflect on each prompt, or write your thoughts down in a notebook or smartphone note:

1. **Write it down.** List three goals, achievements, or desires you recently pursued (at work, home, or in relationships).

The healthiest dopamine system is one that can tolerate uncertainty, take risks, and persist in the absence of immediate results, but also pause and savor when the outcome is achieved.

2. **Dig deeper.** For each of the above, note the following:
    1. What drew you to pursue them?
    2. How did you feel before you achieved or acquired them?
    3. How satisfying was the actual experience? Did pleasure persist, or did craving quickly settle in for "What's next?"
3. **Reflect.** Which patterns of craving do you see? Where does anticipation outweigh contentment? Are any of these goals, achievements, or desires shaped more by others' expectations or cultural narratives than by your own authentic wants?
4. **Start an action plan.** Jot down one action you can take to savor accomplishment—whether it's pausing for reflection, celebrating with others, or simply journaling a moment of gratitude before moving on.
5. **Repeat this exercise weekly.** Over time, you'll notice trends, discover which achievements create sustained satisfaction, and learn how to consciously build habits and goals around actual liking, not just persistent wanting.

Why this matters: Small, intentional self-checks like this build awareness and start to rewire your reward system. Over time, these moments of observation help you make better choices, stop chasing empty rewards, and design a life that blends ambition with real, lasting contentment.

CASE STUDY
Viviana—Achievement Without Satisfaction

Viviana was a project manager admired for her relentless drive and impressive résumé. She had climbed every corporate ladder, led teams, and checked every professional box. Her days were a blur of deadlines, triumphs, and accolades, yet an unexpected numbness began to creep in. No matter how many goals she achieved, the satisfaction faded almost immediately. Even personal milestones—travel, homeownership, relationships—were quickly consumed by restless striving for more.

In coaching sessions, Viviana described feeling like a spectator in her own success, detached from pleasure and confined by invisible expectations. Each promotion or win delivered a brief jolt of excitement, followed by disappointment and the compulsive need to set a new target. "I have everything I told myself I wanted," she confessed, "but none of it makes me happy for long. I'm just . . . stuck on repeat."

We examined Viviana's reward patterns through the lens of neuroscience. Dopamine-driven anticipation kept her locked into relentless pursuit, but the "liking" system—her ability to truly enjoy and integrate accomplishments—was underactive. Chronic exposure to unpredictable outcomes, fleeting recognition, and cultural expectations had rewired her brain to value effort and anticipation—*wanting*—over real pleasure.

Our work focused on breaking the achievement cycle. Viviana committed to scheduled pauses after major successes, allowing herself time to reflect and savor the moment before moving ahead. We built rituals of celebration and acknowledgment, both solo and shared, anchoring the experience of liking alongside the rush of wanting. She learned to identify cultural messages about success and to tune into her personal values, setting goals that reflected authentic desires rather than inherited scripts for what success looks like.

Gradually, Viviana rediscovered satisfaction in a completed project, a quiet evening with friends, and the act of learning for its own sake. Her journey illustrates how understanding the paradox of dopamine can restore joy, resilience, and meaning, even in the busiest and most high-achieving lives.

Shaping Craving Into Contentment: Next Steps

Building lasting fulfillment means working with dopamine, not against it. Awareness is key, and that's the big step you are taking in this chapter. You will dig deeper into working with dopamine and your own relationship with wanting versus liking in Part 3. As you move through those topics, and the remaining chapters of Parts 1 and 2, it will be important to embrace the paradox. It's normal for motivation to surge and satisfaction to fade; you can't change this.

Practical Integration: Your Wanting–Liking Balance Map

One of the fastest ways to embrace the paradox and put this chapter's insights into practice is to track the gap between your own wanting and liking. For two weeks, keep a simple "W/L Journal." Each day, write your answers to these questions:

- What are the top three things you want today? These could be tangible (a dessert, a new purchase) or intangible (recognition from a boss, a message from a friend).
- How do you feel about pursuing them? Note your energy level, focus, and mood during the chase.
- How do you feel afterward? Did the outcome produce lingering satisfaction? For how long? Did it trigger an immediate new craving?
- Did the desire originate from *your* values or cultural pressure and external cues? Do an authenticity check.

After two weeks, review your logs. The patterns are often eye-opening: Many people discover a handful of desires that consistently deliver high liking, and others that reliably fizzle. Once you see these trends, you gain real leverage over your dopamine system. You can deliberately invest more effort into the pursuits that bring both wanting and liking into balance, while reducing time spent on those that leave you empty.

Over time, this exercise rewires awareness into your routine. You'll start noticing in real time when you're on the verge of chasing something that won't leave you feeling satisfied, and you'll have the tools to pivot toward what truly nourishes you. As you move into Part 3, you'll apply these insights to even more targeted practices for rebalancing your dopamine system. Each strategy ahead builds on what you've learned here.

CHAPTER SUMMARY

Throughout this chapter, you explored dopamine's paradoxical role at the heart of motivation and pleasure. Keep these key themes in mind as you move on to Chapter 3 and beyond:

- Dopamine powers intense feelings of wanting and anticipation, but does not guarantee lasting satisfaction; "liking" is less robust in the reward circuit of the brain.
- The treadmill of achievement can lead to chronic craving and muted joy.
- Sex differences and cultural influences shape how you experience desire, drive, and pleasure—it's important to understand your unique "dopamine profile."
- Uncertainty, unpredictability, and risk-taking amplify dopamine, sustaining motivation but risking burnout.
- Real fulfillment comes from integrating wanting and liking—embedding satisfaction into goals, relationships, and everyday practices.

CHAPTER 3

The Pleasure-Pain Balance

Why Too Much Feels Like Too Little

Why do seemingly enjoyable activities—like shopping sprees, streaming binges, gourmet treats, and bursts of digital novelty—so often leave people feeling numb, anxious, or unfulfilled? At the heart of this modern dilemma lies the pleasure–pain balance: the intricate dance by which your brain regulates reward and protects you from excess. What feels good in the moment, if repeated often enough, can quietly tip into feelings of emptiness, dysphoria, and exhaustion. The more you chase the next hit of excitement, the further genuine contentment can seem to slip from reach.

In this chapter, you'll explore how dopamine and associated circuits in your brain evolved to help you navigate scarcity and adversity. (And how it's *not* meant to navigate a world of unending abundance and stimulation.) You'll dig into why your brain is exquisitely sensitive to both pleasure and pain, and why rewards—when they become too frequent, too intense, or too easy to obtain—can set off an internal response that blunts joy, saps motivation, and eventually leads to cycles of burnout and emotional fatigue. Drawing from neuroscience, evolutionary tales, and cross-cultural anthropology, you'll learn how pleasure and pain operate as two sides of the same coin—each shaping the other, and both vital for a resilient, purposeful life. You'll see why pleasure cannot be sustained indefinitely, why overindulgence brings diminishing returns, and why even the most sophisticated brain is designed to seek equilibrium, not ecstasy.

Through the lens of Priya—a young marketing executive who chased every spike of dopamine, only to find herself trapped in growing emptiness—you'll discover just how easily self-reinforcing cycles of pleasure and pain can take hold in modern life. As you read, reflect on your own patterns and also consider sharing your insights with a close friend or loved one. Sometimes the most meaningful change comes from exploring these cycles with another person. Most importantly, you'll gain practical strategies for restoring your own pleasure–pain balance: recognizing the signs of overstimulation, building resilience to modern temptations, and reclaiming a sense of meaning and sustainable motivation, even in a world of limitless excess.

The Pleasure–Pain Seesaw: How the Brain Finds Balance

Beneath every surge of pleasure and every pang of discomfort lies a finely tuned biological system working constantly to keep you in equilibrium. Neuroscientists call it **homeostasis:** the brain's perpetual effort to maintain internal balance despite ever-changing experience.

A CLOSER LOOK
Homeostasis

Your brain constantly monitors internal conditions (temperature, hormone levels, neurotransmitter balance, energy) and makes adjustments to keep everything in equilibrium, or homeostasis. Homeostasis explains why the high after an intense achievement fades, why constant stimulation loses its impact, and why rest and recovery are essential for sustainable motivation and satisfaction.

Every high—an exciting win, a delicious treat, a wave of approval—is matched by an equal and opposite impulse to steady the system. This process is called **opponent regulation**, where your brain activates a counterbalancing mechanism to restore equilibrium whenever it experiences a strong pleasurable stimulus. Think of it like a seesaw: Whenever one side goes up, an equal force pushes the other side down. Say you indulge in a dopamine-spiking treat, such as an impulsive purchase or scrolling on social media. Your reward circuitry surges, and you feel good for a brief time. But your brain is wired to maintain balance, so it triggers counteracting systems (often involving stress hormones or neural circuits associated with discomfort) to bring the system back down.

Dopamine sits at the heart of this balancing act: It spikes dramatically in moments of anticipation and pleasure, helping you notice and pursue something that might help you survive or thrive. But the brain is designed to avoid running these circuits at full tilt for long. When surges of pleasure become frequent or intense—through repeated indulgence, novelty, or overstimulation—the brain recalibrates by reducing the sensitivity of your dopamine receptors, lowering its baseline, or ramping up pain pathways and negative feelings to compensate. By pushing the seesaw too far or too often, the brain's drive to restore balance can tilt you unexpectedly into fatigue, anxiety, or even **emotional blunting**.

A CLOSER LOOK
Emotional Blunting

Emotional blunting is a state where your capacity to feel emotion becomes dampened or numbed. Colors seem less vivid, accomplishments feel hollow, laughter doesn't come easily, and even things that normally matter feel distant or irrelevant. Unlike depression, which involves active suffering, emotional blunting feels more like being behind glass, observing life without fully inhabiting it. It's your brain's protective mechanism in overdrive, dampening all emotional responses to restore balance.

In these cases, the brain has begun prepping for the comedown even before the pleasure ends.

Tracing the Evolutionary Roots of the Pleasure–Pain Seesaw

Evolution set this balancing system in motion as a survival tool. Your ancestors did not expect, or even desire, constant stimulation; pleasure was rare and fleeting, and discomfort frequent. The brief peaks of dopamine kept them motivated to seek out food, shelter, or social bonds. Once a reward was secured, pain mechanisms—fatigue, boredom, or mild discomfort—discouraged overindulgence and brought the system back toward baseline, ensuring time and energy weren't wasted chasing a single source of pleasure. The result was resilience, persistence, and a subtle joy in small, hard-won rewards.

Fast-forward to the present, and the ancient seesaw of pain and pleasure now faces constant swings toward that "high"—luxuries, digital stimulation, instant gratification—that no longer serve survival but instead provoke rebound effects: the **dopamine hangover** of fatigue, irritability, or emptiness.

A CLOSER LOOK
The Dopamine Hangover

A dopamine hangover is the crash that follows intense dopamine spikes. It involves feelings of flatness, low motivation, irritability, shame, or emptiness that can last hours or even days. The greater the spike, the deeper the dopamine hangover—similar to a physical hangover. Many people unconsciously chase the next dopamine hit to escape the hangover feeling, creating a cycle of escalating stimulation followed by worsening crashes.

Nearly every dimension of modern life offers instant access to things that spike dopamine—TV, movie streaming, sweets, social media, online shopping, the list is endless. As a result, your brain's homeostatic systems work overtime, counterbalancing every heightened pleasure with dullness, irritability, or craving for yet another high. If you chase repeated highs without enough downtime, your baseline mood, resilience, and sense of meaning may all drop, even as you seek more to recover those feelings.

Simply understanding the science and evolutionary history of this balance can be liberating in itself. The **opponent process**—that process of opponent regulation and "seesawing" from a high or low back to equilibrium—isn't a flaw. It's a built-in safety mechanism to help you reset, refocus, and adapt. When you become aware of the seesaw's "rules," you can start choosing your pleasures deliberately and embrace life's necessary pains with more perspective, control, and even gratitude.

Practical Strategies: Reclaiming Your Pleasure–Pain Balance

Pause and practice: Before you move on, track how pleasure, pain, and motivation interact in your own life and reveal which habits are supporting or sabotaging your satisfaction. Try this weekly routine:

1. **Explore joy.** List three activities or indulgences you sought out for pleasure this week (e.g., treats, shopping, social media, binge-watching).
2. **Look closer.** For each, note the following: 1. What feelings did you have before the experience (anticipation, craving, restlessness)? 2. How did you feel during the experience? Did pleasure build or fade? Was it deeper with effort or novelty? 3. How did you feel after? Did satisfaction linger, or was there emptiness, fatigue, or a desire for more?
3. **Reflect.** Which activities gave you lasting contentment, and which triggered another cycle of craving or emotional "comedown"? Remember, experiencing a dip after pleasure is biologically normal—a sign your brain is seeking balance, not that you've failed. Pay close attention to which routines rebalance your reward system and rekindle motivation, and which patterns leave you less fulfilled. Notice the difference between activities that restore you and those that deplete you further.

4. **Keep this log for a few weeks.** The patterns will become clear over time. You'll start to see which pursuits genuinely nourish you and which ones just keep up the cycle of craving and comedown.

Why this matters: Building awareness of your pleasure–pain cycles is the first step to breaking out of burnout and emotional blunting. With each small experiment, you reclaim agency over your habits. Eventually, you'll be able to cultivate pleasure that is truly restorative and rediscover the joy that comes not from chasing more, but from savoring what matters. In Part 3, you'll use these insights to deliberately reshape your relationship with pleasure and satisfaction.

The Hidden Costs of Overindulgence: Burnout, Numbness, and Lost Motivation

Issues can arise when you are constantly indulging in desires, ignoring your brain's need for balance. When every craving is within reach and every moment can be filled with stimulation, the line between pleasure and pain blurs—and the toll on your brain, body, and spirit grows harder to ignore. What begins as a search for joy or relief eventually lays the groundwork for burnout, emotional exhaustion, and a creeping sense of emptiness that no new treat or experience seems to fill.

Neuroscience provides a window into this process. Every episode of overindulgence—bingeing on rich foods, scrolling through social media for hours, buying things impulsively, or seeking out repeated thrills—pushes dopamine circuits into high gear. But the brain, built for challenge and scarcity, quickly adapts in self-defense. Dopamine receptors become less responsive, producing smaller bursts of motivation and pleasure from the same activities. What once excited you can soon feel routine or even dull. This phenomenon, known as tolerance or downregulation, affects not only your experience of joy but also your capacity for focus, resilience, and emotional engagement.

The symptoms of overindulgence are often subtle: a sense of fatigue that lingers even after rest, difficulty experiencing gratitude or authentic happiness, irritability, a need for ever greater novelty, and emotional flatness in the face of former passions. Scientists now recognize **anhedonia**—a clinical lack of joy or motivation—as one outcome of chronic overstimulation and link it directly to rates of depression, anxiety, and professional exhaustion. Productivity suffers, relationships erode, and even personal identity can become scrambled by relentless reward seeking and dopamine depletion.

For many, these signs are mistaken for personality flaws, when in reality, they are natural consequences of a brain trying to regain equilibrium after too much sensory input and too many short-term highs. Recognizing the hidden costs of overindulgence is both protective and empowering. By seeing discomfort as a necessary signal (not just something to avoid) and by periodically embracing low-stimulation, effortful routines, you can fortify your dopamine system and rebuild lasting motivation. It's this awareness, and the willingness to confront your impulses with curiosity rather than judgment, that lays the foundation for a healthier, more satisfying life. You've already begun this work in the "Practical Strategies" reflection in Chapter 2, mapping your own patterns of wanting versus liking. In Part 3, you'll take this exploration further with targeted practices designed to rebalance your dopamine system and cultivate lasting fulfillment.

CASE STUDY
Priya—Chasing More, Feeling Less

Priya was a rising star in her marketing firm—social, stylish, always in the know. On paper, she had it all: trendy outfits, the latest gadgets, a seat at every industry event. Her days were measured in dopamine hits—a flash sale, an influencer's post, an unexpected compliment. Yet beneath the surface, she felt a growing sense of emptiness she couldn't shake.

The cycle of indulgence began innocently enough. After a draining week, Priya would indulge: a new pair of shoes, a night of streaming her favorite series, a lavish brunch, or hours scrolling through travel feeds. Each pleasure was intense yet brief, fading faster as her habits scaled up.

Soon, even as the thrills multiplied, satisfaction became elusive; each new treat left her chasing the next, hoping for genuine and enduring joy but never finding it.

At work, Priya's energy flagged. Tasks that once excited her felt mechanical, and creative spark turned to irritation. She found herself glued to her phone, compulsively checking for notifications or "likes"—never truly present, always waiting for a jolt of pleasure that never lasted. Friends noticed her mood changes. "You've got so much going for you, why don't you seem happy anymore?" one asked after a party where Priya barely engaged.

But it wasn't just others who saw that something was wrong. After a restless Sunday night spent staring at her phone, unable to sleep, Priya finally admitted the truth: The more she sought out excitement, the harder it became to feel anything but tired and disconnected. It was her turning point.

In coaching, Priya described feeling "addicted to the rush, but allergic to the aftermath." We explored her routines, tracing the science of dopamine adaptation and the consequences of sustained overstimulation. Step by step, Priya introduced intentional pleasure fasts—weekends offline, brisk morning walks without her phone, simpler routines, and acts of service alongside self-care. At first, these practices felt uncomfortable. Sitting without digital distraction was restless. A walk without checking notifications felt boring. A meal without capturing it for followers felt incomplete. But that very discomfort was the signal she needed. As her nervous system recalibrated, minor pleasures sharpened: A sunset after a hike felt genuinely beautiful, laughter with friends felt energizing, the deep relief of a task completed with patience felt earned. As she broke the loop of relentless seeking, Priya's energy and positive mood returned. The emptiness faded, replaced by a quiet contentment and emotional resilience that no luxury could substitute.

Her journey reflects a universal lesson: More pleasure does not equal more happiness. Only by restoring the pleasure–pain balance—and daring to pause, reflect, and savor—did Priya rediscover the meaning and joy she'd been chasing all along.

Pleasure, Pain, and Resilience: Lessons from Evolution and Anthropology

When anthropologists study ancient and modern cultures around the world, they find that the pleasure–pain balance isn't just a personal phenomenon: It's profoundly shaped by social values, traditions, and ecological realities. How a society defines pleasure, navigates hardship, and builds psychological resilience offers profound lessons for anyone struggling with the costs of overstimulation and burnout today.

In resource-scarce environments where daily survival was never guaranteed, communities often ritualized both pleasure and pain. Celebration after a hunt, a seasonal festival, or communal feasting served to reinforce bonds and mark brief intervals of reward—always followed by time for rest, recuperation, and anticipation of the next challenge. Pleasure was savored in proportion to the work or patience required to obtain it. These cycles of deprivation and abundance taught patience, emotional endurance, and respect for effort, creating a natural protection against excess.

Contrast this with cultures of abundance, whether ancient Rome at its height, or the hyper-connected, consumer-driven societies of today. In these cultures, anthropologists note that rituals and social norms crop up to provide counterbalances against pleasures that are too easily accessed or perpetual: fasting, quiet retreats, creative discipline, shared hardship in sport or story, etc. Research on dopamine and hedonic regulation shows that cultures of abundance create specific conditions—overstimulation, reward saturation, and **hedonic adaptation**—that require deliberate practices of restraint to restore balance and well-being. Even in the most "pleasure-saturated" societies, elders and wisdom-keepers emphasize moderation, delayed gratification, and the importance of adversity as a teacher.

As you learned earlier in this chapter, evolutionary biology shows why this matters. Human brains are designed for cycles, for peaks and valleys, not continuous highs. Individuals and cultures that embrace periods of effort, discomfort, purposeful waiting—and even small doses of healthy pain—tend to be psychologically more resilient, more adaptable, and better able to savor authentic pleasure when it arrives.

Today's world so often encourages escape from discomfort—minimizing pain, maximizing reward, and avoiding challenge. Yet those who build routines that consciously re-introduce healthy friction, "fast" from overstimulation, or seek meaning above mere pleasure develop resilience that cannot be purchased or instantly downloaded. Whether through spiritual practice, physical training, volunteering, or creative discipline, these rituals become the safeguards of joy, helping humans rediscover what feels good.

As you confront your own pleasure–pain balance in the following practice, consider the resources embedded in your heritage: family stories, cultural practices, historical cycles of struggle and renewal. By integrating their lessons and continuing their traditions, you can restore resilience, gratitude, and meaning in ways that counteract the risks of modern abundance.

Practical Strategies: Restoring Balance and Finding Lasting Satisfaction

Pause and practice: This guided, weeklong exercise will help you build habits that replenish satisfaction (rather than drain it) using neuroscience-backed strategies. You don't need to do all of these at once. Your brain responds best to gradual change, so choose one or two practices to focus on this week, and add others over time as they feel manageable. Meet yourself where you are:

1. **Embrace intermittent abstinence.** Pick one stimulating habit—social media scrolling, snacking, online shopping, or streaming—and commit to a twenty-four-hour break. Notice how your urge to check or indulge shifts throughout the day, and how pleasure feels when you return to the activity.

2. **Reframe effort as opportunity.** Choose one task that requires genuine effort: learning a new skill, tackling a creative project, or

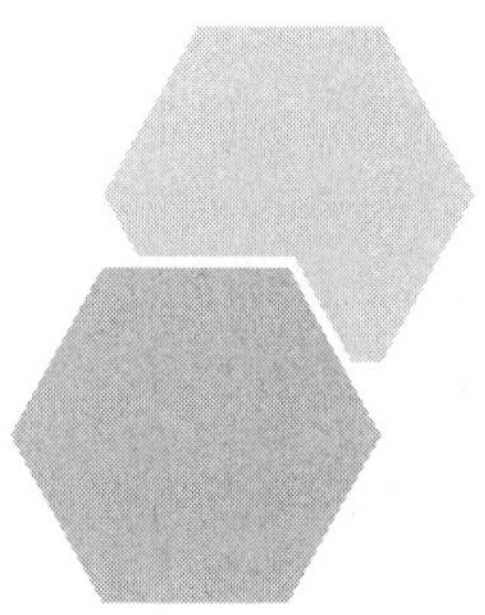

Human brains are designed
for cycles, for peaks and valleys,
not continuous highs.

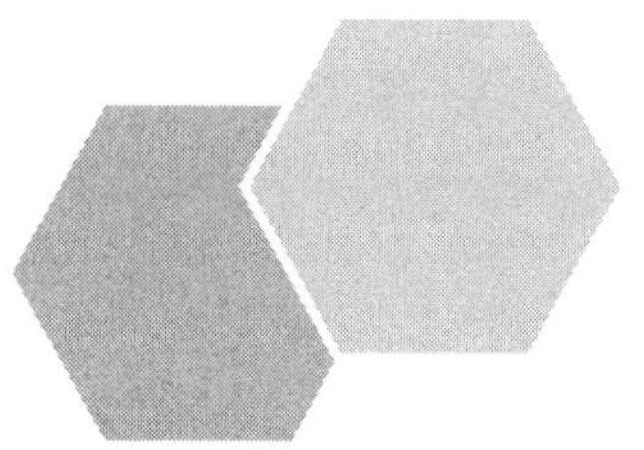

helping someone in need. Work on it for at least thirty minutes during the week. Focus on the process, not just the outcome. Jot down how anticipation and satisfaction build as you persist.

3. **Create an environment of meaning.** Each evening, write down three things you found purposeful or gratifying that day. They can be small wins, moments of connection, or challenges overcome. Practice this gratitude as a way of stabilizing your reward pathways.

4. **Monitor and modulate stimulation.** Designate a device-free zone or time period each day. Replace digital input with physical movement, mindful eating, or quiet reflection. Track how your sleep, energy, and mood shift over the week.

5. **Rediscover ritual and anticipation.** Plan one activity to savor fully: cook a meal from scratch, celebrate a milestone, immerse yourself in music, art, or nature. Let anticipation build, and note the difference in how pleasure feels when paired with patience and intention.

6. **Seek discomfort, then rest.** Challenge yourself physically, intellectually, or creatively. Afterward, schedule proper recovery—whether it's restful sleep, meditation, or unstructured leisure. Observe how effort and rest together boost your well-being and motivation.

7. **Repeat this cycle for several weeks.** Over time, you'll notice which habits reliably replenish satisfaction and which leave you flat or overextended. Adjust your routines to lean into practices that deliver sustainable motivation and joy.

Why this matters: Small, deliberate changes like these help your brain recalibrate, making pleasure deeper and pain more manageable. By working with your biology instead of against it, you transform fleeting highs into ongoing sources of meaning, resilience, and contentment—even in a world overflowing with temptation.

CHAPTER SUMMARY

This chapter unpacked the paradox at the heart of modern reward: The more you chase easy pleasure, the further true satisfaction can slip away. Here are a few other main takeaways from this chapter:

- Your brain's ancient circuitry protects you by restoring balance after every spike of pleasure; this is known as the opponent process.
- Chronic overindulgence blunts genuine satisfaction, leading to emotional exhaustion, burnout, and dwindling motivation.
- Evolution and anthropology reveal that cultures, in addition to individuals, thrive when they honor cycles of effort, restraint, and meaningful recovery—not constant reward.

CHAPTER 4

The Dopamine Myths

Busting Popular Misconceptions

In today's world, dopamine has become the poster child for everything impulsive—blamed for addiction, hailed as the secret ingredient behind motivation, and used to explain why you do what you do. Everyone has been sold the same story: Dopamine is the reason you chase pleasure, scroll endlessly, and struggle to break free from habits that hold you back. But what if most of those headlines were wrong?

This chapter pulls back the curtain on dopamine and exposes the myths that shape how you see yourself and your brain. You'll discover why dopamine is not just a "reward chemical," but a far more complex force that drives movement, learning, and change. When you know the real science, you can spot manipulation in wellness fads, avoid traps set by pop psychology, and make decisions based on how your brain actually works, not how Internet culture says it does. Why does this matter to you? Because every myth about dopamine hijacks your power to shape your habits, regulate your mood, and pursue your goals. Pseudoscience doesn't just confuse—it can limit your potential and keep you stuck. Once you see through these misconceptions, you'll recognize what truly influences your behaviors, and you'll be able to take practical steps to improve your mental health, motivation, and resilience.

This chapter is here to give your understanding of dopamine a reality check. Throughout, you'll confront hype with facts and learn how to protect your mind from trendy solutions that promise too much and deliver too little. You'll learn to let go of the clichés and approach your brain's chemistry—and your own story—with clarity and confidence. You'll also meet Alex, who tried every trendy "dopamine hack" to cure his burnout only to feel more exhausted, showing why quick fixes often fail. It's time to separate noise from knowledge and reclaim the truth that can drive fundamental transformation.

Dopamine Myths: What You've Heard— And What You Need to Know

After decades of working directly with clients and staying at the forefront of neuroscience research, one thing is crystal clear: Misinformation about dopamine is rampant and deeply misleading. Wellness trends, pop psychology, and social media "experts" have taken a complex brain chemical and distilled it into myths that can derail genuine progress. It's time to arm yourself with the real science, so you can break free from confusion and truly understand how to change.

You'll start by digging deeper into common myths surrounding dopamine.

Myth 1: Dopamine Is Just the "Pleasure Chemical"

You've heard that dopamine triggers happiness or pleasure whenever you chase something rewarding. In reality, dopamine's role is about anticipation and motivation—helping you seek, learn, and adapt—not simply delivering a burst of pleasure. As you explored in Chapter 2, the actual sensation of pleasure involves several other chemicals, including endorphins and serotonin, working in concert. Mislabeling dopamine as the sole agent behind joy oversimplifies that intricate neurochemical symphony and leads people to misunderstand their own drives. No reputable neuroscientist would ever claim that pleasure can be boiled down to a single molecule—yet this myth persists.

Myth 2: You Can "Detox" or "Fast" from Dopamine

Popular advice suggests a "dopamine detox"—taking a break from stimulating activities like technology, junk food, or even socializing to reset your brain. Scientifically, you can't turn off or flush out dopamine: It's essential for basic brain function, motivation, and movement every moment you're alive. Calls to "detox" from dopamine are fundamentally misguided—akin to suggesting you can take a vacation from breathing. As an expert, I can assure you that sustainable change comes from recalibrating behaviors, not from magically shutting down vital chemicals in your brain. That said, in Chapter 9, you'll explore what a strategic dopamine reset actually looks like in real practice: not the fantasy of erasing dopamine entirely, but rather the science-backed approach of intentionally reducing high-stimulus environments for a defined period to allow your dopamine receptors to resensitize. This is a very different strategy than the myth of detoxification.

Myth 3: More Dopamine Is Always Better

The idea that boosting dopamine will improve happiness, productivity, or focus is everywhere—but it's misleading. Dopamine levels need to be balanced for optimal function. Excessive dopamine can drive compulsive behavior or contribute to mental health issues like mania, while too little is linked to conditions like Parkinson's disease or depression. The pursuit of "more" for its own sake is not just scientifically baseless—it can be hazardous. Decades of research show that a well-regulated dopamine system is what promotes wellness, not indiscriminate elevation.

Myth 4: Dopamine Is "Good" or "Bad"

It's easy to label dopamine as either a villain responsible for addiction or a hero for motivation. In truth, dopamine is simply a messenger; its impact depends on your environment, habits, and how well other systems (like serotonin and cortisol) are functioning. Demonizing or glorifying dopamine only distracts from the nuances of human experience. If you want to optimize your life, focus on balance and context instead of chasing simple labels that ignore life's complexity.

Myth 5: Dopamine Addiction Is Real

Viral Internet posts claim you can become addicted to dopamine itself, but this isn't scientifically accurate. What people become addicted to are behaviors or substances that over-activate dopamine pathways, such as gambling, junk food, or smartphone notifications. The very notion of dopamine addiction is a distortion of neurochemistry and has no basis in actual brain science. It's important to draw the distinction between mechanism and outcome.

Myth 6: You Can Hack Dopamine Instantly for Motivation

Self-help books and podcasts promise you can spark motivation by "hacking" your dopamine—quick rituals, supplements, or **biohacks** supposedly trigger instant drive. While certain habits and routines do influence dopamine, the process is gradual and rooted in consistent, meaningful action. Sustained motivation results from small changes and attentive self-care, not magical shortcuts. The obsession with instant hacks points to a misunderstanding of how human change and progress truly unfold—it's much more about patience than it is about tricks.

A CLOSER LOOK
Biohacks

Biohacks are behavioral or lifestyle changes designed to use your biology to work in your favor. Examples include cold water immersion to increase dopamine naturally, exercise to elevate mood and motivation, quality sleep to recalibrate reward systems, or time spent in nature to reduce stress and resensitize dopamine receptors. You'll learn more about biohacking in Chapter 15.

Myth 7: Dopamine Controls Everything about Your Behavior or Character

Although dopamine plays a vital role in focus, energy, and drive, it doesn't single-handedly rule your actions or personality. Multiple neurochemicals and brain circuits shape every decision you make and emotional response you have. Assigning dopamine as the sole responsibility for your strengths or shortcomings is not just reductive—it ignores the dazzling complexity of your nervous system. True transformation starts by embracing that complexity, rather than clinging to simplistic explanations.

Dopamine: Stepping Beyond the Myths

By letting go of common dopamine misconceptions, you reclaim agency over your mind and your habits. Free from trendy shortcuts and exaggerated claims, you're able to recognize and work with your brain's natural strengths. When you anchor your efforts in real science, you gain the insight and tools needed to foster meaningful change—shaping focus, drive, and emotional balance from the inside out. This deeper understanding doesn't just remove confusion; it empowers you to build routines and behaviors that endure, setting the stage for lifelong growth and profound fulfillment.

Now that you've uncovered the myths, this chapter will dig deeper into the realities of dopamine.

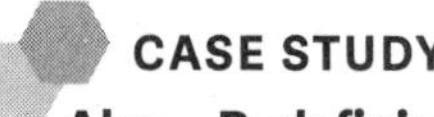

CASE STUDY
Alex—Redefining Success in the Math Classroom

Alex, a middle school math teacher, came to coaching trapped in cycles of burnout and self-doubt. He described days spent tirelessly crafting lesson plans, yet every morning felt like a struggle to summon energy or spark interest in his career. Months of trying "dopamine hacks" from blogs and podcasts—social media detoxes, rigid routines, and motivational playlists—left him more fatigued and frustrated. Alex worried he was losing

Sustained motivation
results from small changes
and attentive self-care,
not magical shortcuts.

his edge: Students seemed disengaged, and creative approaches felt out of reach. The constant stream of advice promising quick resets only deepened his sense of failure.

In early sessions, Alex admitted how overwhelming it was to navigate modern teaching, family life, and personal expectations. He felt guilty for not feeling "motivated enough," believing the myths about dopamine robbing him of resilience. We reframed his goal: Instead of chasing dramatic changes, Alex focused on incremental wins. He began noticing micro-successes: a shy student finally solving a tough equation after weeks of struggle, a spontaneous moment of laughter during a group project when students were debating the best approach to a problem, a thoughtful question from a typically quiet student that sparked genuine curiosity in the whole class, and his own ability to stay present and encouraging even when a lesson didn't go as planned.

We tracked these breakthroughs together, relating them to dopamine's role in reinforcing progress. When setbacks happened (such as a forgotten lesson or a class that went off track), Alex experimented with small replacements, swapping frustration for short moments of reflection or solving a puzzle with students. Instead of seeking perfection, he celebrated persistence and found new meaning in small daily victories.

Gradually, Alex's confidence grew. He brought fresh engagement and patience to his classroom, strengthening routines that supported curiosity and connection. His energy returned—not in dramatic bursts, but as a steady momentum. By honoring the actual science behind dopamine, Alex discovered that fulfillment isn't hacked or forced. Rather, it's cultivated through mindful, incremental change.

Not a Myth: Dopamine and Habits

You've likely heard that good habits (those repeated actions that maintain or improve your health and happiness) are simply the result of willpower, and that enough motivation can break bad habits overnight. In reality, the science of dopamine paints a far more empowering, nuanced picture of lasting change.

It helps to first understand dopamine as the unseen architect behind the countless moments that define your days. It guides your curiosity, sustains your effort, and fuels your growth in ways you rarely notice. Imagine it as your brain's inner compass, quietly orienting you toward what matters, whether that's a spark of excitement over a new idea or the quiet satisfaction of a goal achieved. Think back to those moments when you ventured outside your comfort zone—signing up for a new class, starting a challenging project, or rekindling an old friendship after years apart. Dopamine is working in the background, not simply waving a flag of pleasure, but laying down the neural pathways that turn intentions into actions and fleeting motivation into lasting habits. It is the reason an artist returns to her blank canvas after every setback, and a student finds the grit to master a concept that once felt impossible. In this way, dopamine is less a burst of euphoria and more a silent partner in your everyday life.

What Drives Habit Change?

Over decades of work with clients and immersion in current neuroscience research, I've seen firsthand that lasting transformation isn't about quick fixes—it's about understanding the driving forces behind habits and how they are changed. Here's what the science reveals:

1. **Your brain is equipped with two dopamine-driven learning systems, each located in different regions.** The first system tracks outcomes: It evaluates whether a reward is better or worse than expected (reward prediction error). The second system reinforces repeated behaviors: It tracks whether your actions successfully produce the expected result, using a mechanism called **action prediction error**. Together, these systems shape what you pursue and how you pursue it. The outcome- (or goal-) directed system asks, "Is this working?" while the system responsible for reinforcing repeated behaviors (or habits) automates frequent actions for efficiency. This distinction explains why change can feel so challenging: The goal-directed system requires conscious

effort, evaluation, and mental energy. It's the system you use when you're learning something new, solving a problem, or making a difficult decision. The habit system, by contrast, runs on autopilot. It's fast, effortless, and doesn't require conscious thought. Once a behavior becomes a habit, the brain shifts control from the goal-directed system to the habit system to conserve energy. The habit system doesn't care about your intentions or your goals. It simply executes the learned action automatically. This is why you can intend to skip the morning pastry yet find yourself standing at the coffee shop counter with a croissant in hand before you've even consciously decided. Your habit system fired faster than your goal-directed system could intervene.

2. **Your ability to switch between intentional thinking and automatic action depends on how flexible your brain networks are and how well your body manages dopamine.** That's why science-backed "replacement" strategies—substituting new, healthier actions for outdated habits—are genuinely effective. When you swap scrolling your social media late at night for a new, meaningful ritual, such as a **somatic** practice, you're using both the goal-directed system and the habit system together. The goal-directed system helps you make the conscious choice to replace the old behavior, while repeated practice encodes the new behavior into the habit system, making it automatic over time. This is how deep, sustainable change happens.

3. **Dopamine works with surgical precision, only activating the neural "hot spots" relevant for specific moments and behaviors.** Every time you persist through a challenge or repeat a rewarding routine, dopamine quietly strengthens those specific neural pathways, making the next effort slightly easier. Understanding this empowers you to focus on the rituals and repetitions that build momentum, rather than fixating on onetime "quick fixes."

4. **Lasting change doesn't hinge on sweeping gestures or empty "hacks."** Neuroscience confirms that authentic transformation emerges from countless, small, repeated choices—each act carving

new pathways in your brain. Whether you are dealing with challenges related to concentration, mood, or motivation, focusing on habit change can reshape your neural landscape and fuel true progress. This science is empowering: No matter your starting point, resilience and growth are not only possible—they're hardwired into your biology. Success comes from valuing progress, not perfection.

⌕ A CLOSER LOOK
Somatic

Somatic refers to anything related to the body or bodily sensations. In this context, somatic practices engage your body's physical sensations and awareness intentionally, bringing attention to what you're feeling in your muscles, skin, and nervous system rather than focusing on your thoughts or external stimuli.

Even if you sometimes feel stuck, distracted, or defeated, habit change is always within your reach. Your brain is engineered to learn from repetition, adapt with feedback, and celebrate every small step forward. By focusing on the journey, appreciating even minor improvements, and tying your daily routines to personal meaning, you activate the power of dopamine. Growth becomes accessible; setbacks become opportunities to adjust and grow. The secret isn't found in shortcut "hacks" or fleeting trends—it's in honoring the elegant science within every cell of your brain. As you work with this knowledge, you move beyond fleeting bursts of change into progress that truly lasts.

Practical Strategies: Applying Dopamine Science for Change

Pause and practice: Take a pause here to do some work in what you've learned about dopamine throughout this chapter. Based on neuroscience, the following approaches will help you shape habits and behaviors with real staying power. Rather than attempting all of these strategies at once, pick one approach that resonates with your life, practice it consistently for two to three weeks, then layer in a second strategy. The brain responds most effectively to small, incremental changes.

1. **Identify and redesign your habit cues.** Notice the triggers that set your routines in motion, whether it's a time of day, an emotional state, or a specific situation. To break or build habits, adjust these cues: Move your phone out of sight before sleep, to remove the visual trigger that activates the scrolling habit, set visual reminders for new goals, or schedule productive tasks when dopamine levels are naturally higher (in the morning or after light exercise).

2. **Anchor new habits to existing routines.** Pair new behaviors with well-established ones. For example, review your goals while sipping morning coffee or thank a colleague immediately after a meeting. This leverages your brain's penchant for pattern recognition, making change feel more natural.

3. **Track and celebrate incremental progress.** Keep a log of tiny wins—finished tasks, breakthroughs, or even attempts that didn't go perfectly. Acknowledge and reward these moments; frequent, accurate feedback stimulates dopamine's reinforcement pathways and boosts motivation.

4. **Use the power of replacement rather than restriction.** Rather than trying to eliminate a destructive habit in isolation, substitute it with a meaningful alternative. Replace evening mindless scrolling with reading or reflection. Replacement signals leverage both goal-directed and habitual circuits—making change stick.

5. **Design supportive environments.** Arrange your spaces so desired habits are easy and visible: Place healthy snacks within reach; create a clutter-free workspace for focus. The brain responds to context—make your environment a silent partner in your growth.

6. **Seek social accountability and encouragement.** Share your intentions and successes with supportive friends, family, or peer groups. Celebrating victories together and receiving feedback primes dopamine's social reward circuits, helping new habits endure.

Why this matters: Building lasting change isn't about perfection. By repeatedly applying these practical strategies, you'll harness dopamine's real power—turning science into progress you can see and feel in everyday life.

CHAPTER SUMMARY

By discrediting the myths and uncovering the truths about dopamine, you've gained practical tools to reshape how you think, feel, and act. Unlocking dopamine's secrets equips you to move forward with confidence, building habits and outcomes that are both resilient and truly rewarding. Here are the key takeaways from this chapter:

- Your brain doesn't heal through deprivation; it heals through intelligent design. The most effective approach involves redirecting dopamine's motivational power toward meaningful behaviors and away from **high-variable-reward traps** (activities like social media scrolling, slot machines, or binge-watching, where rewards arrive unpredictably and keep you hooked through uncertainty rather than genuine satisfaction).
- Dopamine is a silent partner in your daily life, driving motivation, learning, and choices.
- Your mind uses two complementary systems: one tracks outcomes (evaluating whether rewards meet expectations), and the other reinforces repeated behaviors (strengthening actions that successfully produce results).
- Using these systems to make minor adjustments to cues, environments, and daily rhythms creates powerful, lasting shifts in behavior.
- Substituting refined habits for old ones unlocks progress and resilience—step by step, not all at once.

Under the Influence

Navigating a World
That Shapes Your Brain

In Part 1, you uncovered the science behind dopamine and the natural systems at work in your brain. But there are also external forces at work, pushing to influence these systems and hijacking your dopamine receptors. Social media, news alerts, shopping platforms, digital games, and the constant hum of technology are all competing every day for your attention, shaping the very contours of your thoughts and feelings. In Part 2, you'll gain a clear, practical understanding of how these influences work, why they're so powerful, and which steps you can take to reclaim your focus, motivation, and authentic connection.

Across the next three chapters, you'll break down the "dopamine industrial complex"—the business of engineered digital addiction—and see precisely how companies use targeted design, habit-forming features, and persuasive content to keep you scrolling, buying, and craving. You'll learn why "just five more minutes" can quickly turn into an hour lost to a dopamine-fueled loop, and how digital rewards compete with real-world goals and relationships for space in your mind. You will also discover how to map personal dopamine triggers, habits, and experiences onto a dynamic selection of "menu items": a curated set of activities, rituals, and practices that provide healthy reward without hijacking your system. You'll craft rituals from these options that not only spark immediate reward but also prevent burnout and ensure vitality through life's ups and downs. The chapters here guide you through step-by-step strategies for self-experimentation with which activities satisfy your brain, a menu audit of your daily choices, and ongoing refinement of these choices.

You'll also explore the paradox of loneliness in an age where everyone is constantly connected. Through current neuroscience and real-world examples, you'll discover how the search for digital validation rewires your brain's reward circuitry, fueling the fear of missing out as well as comparison and shallow connections. The tools and skills in this part will help you as you start building satisfying relationships and lasting resilience in Part 3.

Finally, you'll dive deep into the effects of trauma on dopamine; you'll learn how painful experiences can disrupt motivation, deepen emotional numbness, or spark compulsive behaviors that seem beyond your control. These chapters utilize clinical insight, grounded case studies, and actionable strategies to help you transition from feeling stuck to developing healthy, self-driven habits and emotional balance. By the end of Part 2, you'll recognize the things that keep you distracted, disconnected, or discouraged. This isn't just about surviving the digital age; it's about thriving on your own terms, equipped with practical knowledge and proven tools for genuine well-being.

CHAPTER 5

The Dopamine Industrial Complex

How Tech, Media, and Markets Hijack Your Brain

You live in a world that's mastered the art of keeping you hooked. Every day, the apps you open and the notifications you receive are optimized by billion-dollar industries to grab your attention, trigger your cravings, and nudge your behavior in ways that feel almost invisible but become deeply ingrained.

This chapter uncovers the neuroscience, psychology, and business behind "engineered addiction," revealing how social media, video games, streaming, and online shopping exploit your brain's dopamine circuits. You'll learn how these technologies don't just mirror ancient human drives for novelty, status, and social connection—they amplify and manipulate them with algorithms tuned to reward you at just the correct interval, for just the right action, until self-control feels like a struggle you're destined to lose. Through a unique blend of scientific research and sociological insight, you'll uncover how your most primal motivational systems have been repurposed for profit in ways evolution never prepared you for.

But this isn't just about the tech companies. You'll gain a clearer view of your own brain's vulnerabilities: why some digital rewards become irresistible, why intermittent "likes" or flash sales are harder to walk away from than you realize, and why there's often a fine line between engagement and compulsion. You'll also meet Eli, a paralegal whose quest for escape became a nightly ritual, unraveling the hidden costs of life in a world engineered for your dopamine. By the end of this chapter, you'll have the tools to spot engineered influence, reclaim control, and choose how you engage—empowering you to navigate the digital world on your own terms and reconnect with what truly matters.

Engineered Addiction: The Science and Evolution

From the dawn of human civilization, survival hinged on tracking patterns, responding to novelty, and connecting within groups. Primitive drives—like the rush of anticipation before a successful hunt or the satisfaction of being valued in a tribe—were orchestrated by the same dopamine pathways that drive your tastes and decisions today. What's radically changed isn't your biology, but the environment engineered around you.

Every ping, alert, and scroll is built on that ancient reward circuitry. As you explored in Part 1, dopamine spikes helped your ancestors stay motivated, learn from surprises, and adapt. Today, these exact neural mechanisms are tapped and amplified by technologies and platforms specifically designed to maximize engagement. Users can end up drawn into hours of interaction, often without clear intention or genuine fulfillment.

On a neuroscientific level, when something exceptional or unanticipated happens—a rare berry discovered in the wild or a sudden burst of digital likes—your brain releases dopamine to reinforce the behaviors that led to that outcome. Platforms exploit this by using reward prediction error (see Part 1) and **variable ratio reinforcement**, keeping your brain guessing and coming back for more.

🔍 A CLOSER LOOK
Variable Ratio Reinforcement

A principle borrowed from classic behavioral psychology and studies of animal learning, variable ratio reinforcement is a pattern of reward delivery where you don't know exactly when the next reward will come, but you know it will come eventually. Instead of a predictable reward every time (you scroll once and get a notification), rewards arrive unpredictably after a varying number of actions (sometimes after two scrolls, sometimes after fifteen).

When digital platforms mix predictable rewards (a notification, a message, a sale) with unpredictable ones ("You won't believe who liked your post!"), the dopamine system fires at its strongest. These erratic patterns mimic slot machines, where every spin could bring a reward, keeping engagement levels high and prompting users to return even after brief periods of disengagement. In evolutionary terms, this kind of uncertainty kept humans motivated to explore and forage even when immediate rewards were scarce. But in the digital age, it means simply checking your phone or inbox can quickly turn into a compulsion (also called the **digital compulsion**).

🔍 A CLOSER LOOK
Digital Compulsion

Digital compulsion is an irresistible urge to check your phone or scroll through social media, even when you don't actually want to. Unlike casual habit, it feels mandatory because variable ratio reinforcement has rewired your dopamine system to treat these devices as essential. Your brain perceives the urge similar to how it perceives physical hunger or thirst.

The act of checking, even when you know it might yield nothing new, feels worthy because sometimes it does result in unexpected pleasure: a message, a like, a flash sale, a breaking headline. It's not weak willpower; it's an exquisitely sensitive feedback loop designed to make it difficult to disengage from technology.

Psychologically, these digital rewards serve as new sources of validation and belonging. Updates and images become benchmarks for worth, while the uncertainty about how you'll be received ("Will I be liked?" or "Will this comment be noticed?") generates a strong emotional tension within. This engineered unpredictability activates the brain's "seeking" state, making the appeal of digital interaction far greater than the value of any individual click or message.

Sociologically, persistent connectivity and curated digital environments reshape what people crave. The desire for social recognition, status, and personal achievement—once satisfied through real-world accomplishments and communal celebration—is now measured in fleeting metrics, such as "likes," and reciprocal exchanges, such as shares and viral trends. Cultures internalize these new standards quickly, and more aspects of life continue to move toward the algorithms of advancing technology.

Crucial to these shifts is the fact that the dopamine system is adaptive. As rewards become predictable, they lose potency, driving users to pursue greater novelty or intensity—a phenomenon seen in everything from social feeds to gambling platforms to binge-watching TV series. This constant escalation doesn't lead to deeper fulfillment but instead to some hefty psychological, social, and even financial costs. This shifting threshold, a neurobiological phenomenon called **hedonic adaptation**, quietly erodes resilience and motivation, making it harder to persist with anything lacking instant payoff (like that TV binge, etc.). You'll learn more about the impacts of media on your well-being later in this chapter.

Fortunately, understanding the brain's evolutionary wiring, the psychological principles being manipulated, and the sociological shifts underway can empower you to examine your relationship with technology with greater clarity and compassion. You're navigating environments that weren't part of your ancestral landscape, and recognizing this is the first step toward reclaiming agency and redefining what genuine reward and connection mean for you.

A CLOSER LOOK
Hedonic Adaptation

Hedonic adaptation is why you increasingly need stimulating content, shopping, or digital engagement to feel the same satisfaction. Understanding hedonic adaptation explains why bigger and better is never enough and why replacement behaviors work better than chasing the same rewards endlessly.

How Algorithms, Media, and Markets Use Personalization to Hijack Dopamine

Every time you open an app, browse a shopping site, or even watch a streaming platform's preview, you're entering a carefully constructed ecosystem built to monitor, predict, and manipulate your behavior. Underneath the polished interfaces and personalized options, lines of code steadily collect data points—what you glance at, click, skip, or linger on. In a split second, **algorithms** adjust your media experience, presenting you with new hooks uniquely tailored to your personal dopamine circuits.

A CLOSER LOOK
Algorithms

An algorithm is a set of automated instructions that tells a computer how to process data and make decisions.

Unlike traditional advertising, which, in the past, had casted a wide net to capture your interest, today's digital platforms deliver "hyper-personalization" in real time. This means the system learns, through reinforcement and trial and error, exactly which images, headlines, notifications, and offers are most likely to make your brain release dopamine and compel action. Once an algorithm discovers a product or type of content that hooks your attention, it quickly ramps up the exposure—and with every interaction, its predictions become sharper. This is why simply pausing a video can soon lead to a feed filled with similar content, or a single search for travel boots triggers weeks of targeted ads and notifications.

Escaping the Cycle: What Digital Compulsion Costs You

You might notice it creeping in on quiet evenings—a vague sense of emptiness when the notifications slow, or a frustrated restlessness as you swipe from app to app, never quite satisfied. Over time, the cycle of digital engagement begins to spill into every corner of your life, disrupting not only how you spend your time but also how you think, feel, and relate to others. It can even cost you a good deal of money.

Psychological Costs

The psychological impacts of digital compulsion are profound. On the surface, devices like smartphones and TVs foster excitement and novelty. Dig deeper though, and you'll find the cost: cognitive fatigue, reduced focus, and a growing intolerance for boredom or stillness. The constant stream of dopamine-triggering experiences recalibrates the brain's sense of "normal," making tasks that don't offer instant feedback or rewards—like reading, **deep work**, or face-to-face conversations—seem less appealing or even aversive. It becomes difficult to find pleasure in simple, real-life moments.

A CLOSER LOOK
Deep Work

Deep work is sustained, focused mental effort on mentally demanding tasks. It's the kind of concentration required for complex problem-solving, creative thinking, writing, analysis, or skill development. Deep work demands that your prefrontal cortex (the planning and focus center of your brain) remains activated while your dopamine system stays patient, resisting the pull toward quick rewards and novelty.

Over time, this chronic overstimulation leads to heightened stress, sleep disruption, diminished physical health, and even an inability to regulate your emotions. Indeed, for some, media can become a way of sidestepping uncomfortable emotions. Boredom, stress, and more get pushed away and numbed by escaping into a device. Eventually, persistent avoidance of emotions through digital compulsion chips away at a person's overall resilience, making it harder to engage with tasks that require sustained focus or emotional investment. In client after client, I've witnessed higher levels of anxiety, irritability, and even mild depression when the draw of screens repeatedly supersedes the need for proper rest or meaningful in-person connection.

This isn't just anecdotal. Across age groups, clients report feeling less capable of tolerating quiet, undistracted moments; the baseline for excitement shifts, so that everyday life seems dull compared to the drama of the Internet. What emerges is not merely a personal challenge, but a reflection of how today's technologies have recast the emotional landscape. The feedback loops of digital platforms make it all too easy to mistake stimulation for satisfaction—leaving people feeling more disconnected from themselves than ever before.

Social Costs

On a sociological level, digital media transforms not just individual habits, but collective expectations about communication, aspiration, and self-worth. Algorithms portray viral content, trending topics, and social metrics (follower counts, like tallies, etc.) as public measures of value. This shifts communities toward a never-ending chase for validation, leading not just to individual anxiety but to new forms of group pressure and status anxiety. Social media doesn't just connect you—it compares you and encourages you to compare yourself. Follower counts, reactions, and story views all offer tantalizing but fleeting glimpses of belonging. Several studies (including "The Effects of Social Feedback Through the 'Like' Feature on Brain Activity: A Systematic Review Using fMRI and EEG," published in *PubMed Central*) have shown that digital stand-ins for connection create a persistent hunger for more, which standard social interactions often can't match. As a result, even casual use of media can spiral into compulsive cycles, where chasing digital rewards gradually overshadows the satisfaction of real connection or mindful presence.

Compounding these social effects are other subtle pressures: When everyone else seems constantly available, quick responses become the norm. The expectation to always be online, to reply quickly, or to participate in group threads can make stepping away feel risky, even if what you gain from these connections is minimal. It's no accident that many apps have features that highlight your absence—"last seen" markers, read receipts, or auto-alerts when you haven't checked in—encouraging anxiety and nudging you to reengage, even if it's against your better judgment. (You'll discover more about the impacts of constant availability in Chapter 6.)

Research in digital sociology and behavioral neuroscience shows that this persistent connection rarely leads to true satisfaction. Instead, it can erode self-esteem; comparing your life to the carefully curated digital lives of others sets unrealistic expectations for your own happiness, success, or attractiveness. Just a click, and you are surrounded by images and captions telling you what you are missing or how you don't measure up. The relentless stream of updates also produces **ambient stress**, a subtle but measurable tension experienced as your mind toggles between feelings of anticipation and unease.

A CLOSER LOOK
Ambient Stress

Ambient stress is a low-level, persistent anxiety or tension that sits in the background of your life. It's not a crisis or an acute threat, but rather the chronic hum of worry or pressure that never fully resolves.

The result? A world where unplugging feels unnatural—not because it actually is, but because your brain, shaped by both ancient wiring and modern engineering, struggles to discern real connection and fulfillment amid an endless parade of artificial rewards.

Financial Costs

The economic impact may feel subtle, but it does exist. Online shopping platforms leverage dopamine-triggering tactics—flash sales, limited-time offers, one-click purchases—to make money on the brain's urges. These engineered temptations can override thoughtful decision-making, leading to overspending or guilt-laden "retail therapy" sessions. The intersection of convenience, persuasive marketing, and neuroscience means that willpower alone is often not enough to curb repeated impulse purchases.

Looking Beyond the Screen: You Aren't Alone

As personal as the struggle with digital compulsion may feel, it is one that so many experience. Over the last decade, I've worked with countless clients (such as Eli, whose story you will explore in the next section of this chapter), grappling with the effects of nonstop digital stimulation on their mood, sense of purpose, and relationships. Time and again, I've heard descriptions of life online as a swing between bursts of excitement and stretches of emotional flatness. Many recall the fleeting rush of affirmation—a new like, a message, a clever video—followed almost immediately by a dip in satisfaction, prompting them to seek the next hit, whether by checking their messages, refreshing news feeds, or diving into another round of gaming.

Recognizing the costs of digital compulsion is not a cause for self-judgment—it's an invitation to reclaim agency. By shining a light on these influences, you give yourself permission to pause, reflect, and chart a new course—one that prioritizes your well-being over platforms' profits, and puts you, not the algorithm, back in charge of your attention and aspirations. You will begin this work in the "Practical Strategies" activity in this chapter and continue it in Part 3.

CASE STUDY
Eli—The Digital Loop

Eli, a paralegal at a corporate law firm, first noticed that his evenings began feeling strangely incomplete unless it involved multiple screens—the TV, his phone, etc. What started as a quick check of social media, a round of gaming, and a glance at shopping deals gradually evolved into a nightly routine, a private world apart from work and family, where stimulation was always within reach. At first, these moments felt rewarding: new messages, unexpected victories in games, and the thrill of finding an exclusive online offer all provided a sense of gratification.

But the cycle quickly intensified. On the nights when notifications were few and victories came slowly, Eli felt a mounting sense of agitation. He could no longer unwind without his digital rituals, and his sleep grew restless and irregular. He realized that hours had disappeared at night while he was using media—a fog settling over his focus and energy the next day. His relationships began to suffer. Dinner with friends felt like a distraction from checking in on his digital life, and real-world interactions could not compete with the instant gratification of online rewards.

Over time, Eli noticed that the emotional highs were less frequent, replaced by a nagging urge that he rarely found true satisfaction in. The stress relief and excitement he once felt gave way to irritability and a sense of emptiness. Attempts to take breaks or limit screen time increased his discomfort, and he often returned to old habits to escape that restlessness. What started as simple leisure became a way to avoid difficult emotions—from boredom to stress and more—creating a feedback loop that shaped his routine and outlook.

In coaching, we started by mapping Eli's digital triggers: the specific times, emotions, and situations that pulled him toward screens. Instead of attempting cold-turkey elimination (which research shows backfires), we redesigned his evening menu. We identified one or two low-dopamine but genuinely satisfying activities that could replace the multiscreen spiral: a twenty-minute walk after work, time spent building something with his hands, and intentional conversation with his partner without devices present. We also adjusted his environment by setting his phone in another room during dinner and morning hours, removing the visual cue that triggered the habit cascade.

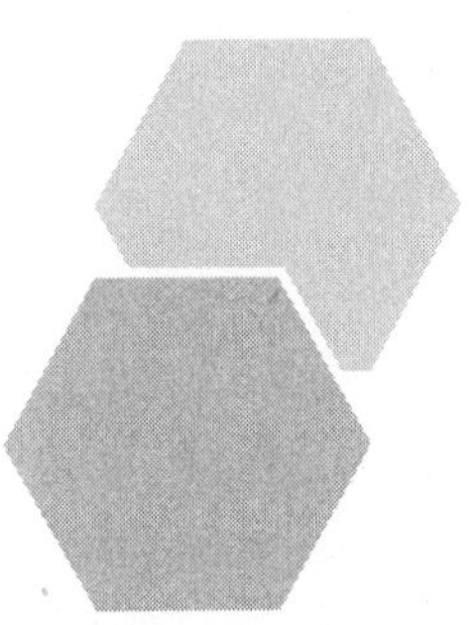

You're navigating environments that weren't part of your ancestral landscape, and recognizing this is the first step toward reclaiming agency and redefining what genuine reward and connection mean for you.

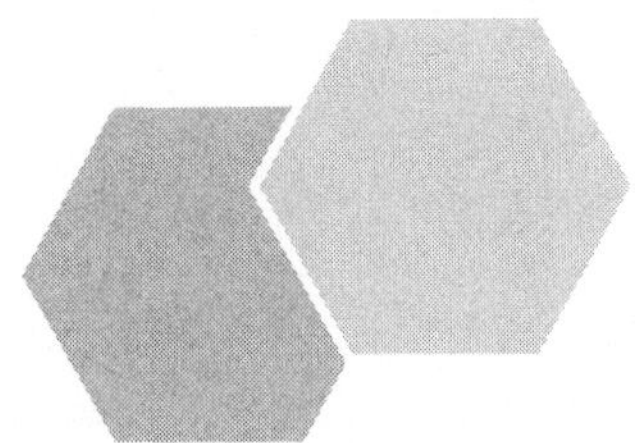

Critically, Eli tracked the subtle changes: how his sleep improved within two weeks, how his focus at work sharpened, and how real conversations began to feel rewarding again rather than like distractions. These small wins reinforced the new habit circuit. By addressing the actual dopamine dysregulation rather than relying on willpower alone, Eli gradually reclaimed his evenings and his presence.

Eli's story is far from unique. The engineered environments behind social platforms, games, and e-commerce are designed to engage and hold attention, leveraging dopamine circuits to make it increasingly difficult to disengage. Breaking the cycle required more than willpower; it meant understanding—and undoing—the systems that made these habits so powerful became the path to genuine freedom.

Practical Strategies: Outsmarting Engineered Digital Addiction

Pause and practice: When you understand the causes of digital compulsion and how digital platforms work to draw you in, you're better equipped to choose your digital participation intentionally. You can learn to separate engineered friction from authentic motivation and begin building patterns that support your actual well-being, not just the fleeting stimulation of a dopamine hit. The following are many strategies I use with my clients to help them reclaim control over their attention, habits, and well-being. These proven approaches will support you, even in a world crowded with carefully engineered digital temptations.

1. **Audit and interrupt your autopilot.** Begin by tracking your screen time and digital routines for a week. Set intentional times to unplug—like device-free dinners or a tech pause after work—to break up compulsive engagement and notice what triggers your use.

2. **Redesign your notification landscape.** Turn off nonessential alerts and prioritize only the apps that genuinely improve your life or well-being. Move distracting apps to harder-to-reach folders and set specific windows for checking messages or social media.

3. **Swap instant gratification for meaningful alternatives.** Rather than simply restricting use, replace scrolling or shopping rituals with restorative activities: take a walk, read, meditate, or connect with a friend. The more you substitute healthy rewards, the easier it becomes for your brain to recalibrate motivation.
4. **Create device-free zones and rituals.** Designate specific spaces— like the bedroom or dining table—as digital-free. Anchor enjoyable offline habits, like journaling or quiet hobbies, to these areas to retrain your reward circuits for non-digital sources of satisfaction.
5. **Schedule real-world connection.** Commit to regular plans for in-person social time, whether that's coffee with a friend, a class, or volunteering. Genuine face-to-face engagement naturally boosts dopamine and counteracts the stress induced by screen-based habits.
6. **Use accountability and social support.** Share your intentions with trusted friends, family, or support groups. Celebrating small wins and setbacks together taps into social reinforcement and builds lasting motivation to change.

Why this matters: By consistently applying these practical strategies, you'll reshape your digital routines and strengthen attention, connection, and fulfillment—using the neuroscience of dopamine to guide progress you can see and feel every day. In Part 3, you will continue practices to boost focus, connection, and joy in this hyper-digital world.

CHAPTER SUMMARY

Understanding how tech, media, and markets target your dopamine system gives you the knowledge to reclaim your time, attention, and connection. Here are the main takeaways from this chapter:

- Digital platforms and algorithms exploit natural reward circuits, making compulsive engagement (called digital compulsion) far more likely.

- Engineered uncertainty and intermittent rewards strengthen habit loops, fueling distraction and reducing satisfaction with everyday life.
- The impact of digital compulsion can erode creativity, focus, emotional well-being, relationships, finances, and more.
- Practical strategies like interrupting autopilot routines, redesigning your notification landscape, and prioritizing real-world connections can reshape your digital habits.

CHAPTER 6

Dopamine and the Disconnected Self

Loneliness in a Hyperconnected World

You live in an era where connection is promised everywhere—video calls, group chats, virtual communities, and a scrolling feed of lives unfolding in real time. Yet for countless people, these digital ties feel paper thin, failing to fill the more profound human need for genuine, felt belonging. You can be "always on" yet rarely "fully seen."

This chapter digs deeper into the gap between surface-level connection and persistent feelings of loneliness, with a special focus on the underlying chemistry: dopamine. Grounded in neuroscience, you'll see how technology platforms designed to unite people actually exploit the dopamine system, pushing you to patterns of craving and validation that intensify isolation instead of easing it. You'll explore how modern technology reconfigures relationships, turning genuine social bonds into cycles of reward. Likes, notifications, and online approval trigger quick bursts of dopamine but rarely soothe the longing for true connection. Over time, the attachment system confuses availability with intimacy, leaving people socially busy yet emotionally unfulfilled at the same time. This in turn leads to a chase of fleeting social rewards instead of lasting belonging.

As you read on, you'll learn why loneliness is on the rise, how technology shapes dopamine and emotional health, and what science-backed steps can help redirect the reward system toward real connection. You'll uncover the isolation felt by Anna, who found a sense of belonging increasingly out of reach despite being constantly "connected" online. This chapter is not just an investigation of solitude, but a guide to understanding and overcoming the growing issue of loneliness. You'll find simple, practical experiments—phone-free meals, more extended conversations, and purpose-driven groups—that reawaken reward through meaningful human presence. The aim is to feel truly connected, not merely connected to, in daily life.

The Roots of Loneliness: An Ancient Signal in a New World

Loneliness is not a modern affliction; it is an ancient signal embedded in the human brain by millennia of evolution. Early humans depended on their tribe for survival in a world filled with predators, harsh elements, and uncertain resources. The cost of living alone could be death. So, the feeling of loneliness evolved as an adaptive warning—powered not just by emotion, but by dopamine-driven motivation. This chemical reaction created urgency: Just as pain steers the hand away from fire, dopamine and discomfort pulled people toward connection, comfort, and safety.

Today, that exact evolutionary mechanism remains active. When people feel socially excluded or perpetually alone, the brain's threat detection and pain centers light up. However, this mechanism has been hijacked by digital media. Dopamine, which should push people toward rewarding connection, now often fuels restless searching for approval, recognition, or for some signal that the next virtual interaction will finally provide relief. In this hyperconnected world, dopamine's system for resolving loneliness is triggered again and again, without leading to true fulfillment.

In the next section, you'll learn more about how the evolutionary urge for connection goes unsatisfied by the shallow rewards of the Internet.

The Digital Reward Circuit: How Screens and Social Chemistry Hijack Your Dopamine

Modern technology has offered unprecedented opportunities for social contact. However, beneath the surface, research into the effects of digital life reveals profound disruptions in the brain's social circuitry. Face-to-face interaction works through many senses at once—eye contact, tone of voice, touch, and body language—all powered by neurochemicals like oxytocin and dopamine. These chemicals help build trust, emotional resonance, and the sense of safety needed for someone to feel truly understood in a relationship. In digital spaces, these signals are weaker, slower, and often shallow. Messages, likes, and notification pings deliver instant bursts of dopamine that stimulate the reward system, but they rarely activate the oxytocin response that builds real intimacy.

Over time, too much screen exposure can subtly change how the brain experiences social rewards. Brain scans show that social media engagement can alter how different regions in the brain—namely those responsible for emotional regulation, perception, and connection—respond. The ventral striatum, which is central to the reward system, may respond strongly to digital feedback, but less so to the nuances of face-to-face interactions. Meanwhile, the **insula** and **anterior cingulate cortex**—regions in the brain that shape empathy and emotional self-awareness—tend to be less active in lonely individuals who spend more time online. This means the digital environment, for all its stimulation, can leave users less tuned into the emotional depth of real-life encounters.

Worse, screen-based connection relies heavily on **mentalizing**, which is the ability to infer mental states such as beliefs, desires, and intentions in ourselves and others. The brain works overtime to decipher the deeper meanings of texts, emojis, and comments instead of reading facial expressions, tone, or context. The result is more mental effort with less emotional reward: digital connection feels busier but far less nourishing. Intimate conversations lose their depth; group interactions get broken up by constant distractions and comparisons.

○ A CLOSER LOOK
The Insula and Anterior Cingulate Cortex

The insula: Processes internal body signals (like heart rate, breathing, and gut sensations) and links them to emotional experience. It's also central to empathy, helping you recognize and respond to others' emotional states. When the insula is less active, people struggle to tune into their own emotions and the emotions of those around them.

The Anterior Cingulate Cortex (ACC): Responsible for emotional regulation and self-awareness. It helps you notice when something feels wrong emotionally, resolve internal conflict, and adjust your behavior based on social feedback. When the ACC is underactive, people often miss emotional cues from others, feel disconnected from their own feelings, and struggle to adapt their behavior in social settings.

Eventually, constant online engagement can rewire the brain's reward system. People may be more alert to social threats but feel less joy from real connection. The brain chemicals meant to foster social adaptation, such as oxytocin and dopamine, are rerouted into a feedback loop of seeking approval and mistaking screen time for genuine closeness. Reclaiming the chemistry of real connection starts with noticing these subtle changes—and by creating spaces that support both deep trust and emotional resonance, far beyond screens and notification.

○ A CLOSER LOOK
Mentalizing

Also called the theory of mind, mentalizing involves understanding that other people have beliefs, desires, and intentions that may differ from our own, and that their behavior stems from these internal states, not just external circumstances.

Recognition versus Resonance

Recognition and **resonance** are terms that can be used to describe what is missing in digital interactions. Recognition is defined as the acknowledgment of your existence through external metrics: a quick, impersonal form of validation that offers a brief dopamine hit but rarely satisfies. Examples include likes and follower counts, which offer a quick boost of dopamine but rarely linger as true satisfaction. In contrast, resonance is the feeling of being understood and accepted by trusted others; this feeling sends sustained waves of dopamine and oxytocin through the brain. Each moment of resonance—whether across a table or in shared silence—anchors the reward system where it belongs: not in a notification, but in the steady presence of authentic bonds. Dopamine becomes not a driver of loneliness but a signal that satisfaction has found its home.

Constant Availability: The Hidden Costs

As you read in Chapter 5, one by-product of social media is a sense of constant availability. A person can be reachable all day, and many have come to expect that others *will* be available to them at any time. The anticipation of rapid replies, constant updates, and seamless participation in group chats has turned the old rhythms of social life into fragmented, performative exchanges. Users become conditioned to equate availability with closeness, while genuine intimacy quietly erodes. A peculiar loneliness is left in its place: One can be *reachable*, yet rarely reached in a *meaningful way*.

Neuroscience shows that this state of being constantly distracted by phones and digital alerts makes it harder for the brain to process the nuance of different emotions, focus on relationships, and form lasting memories. The prefrontal cortex, tasked with regulating focus and impulse, gets overwhelmed by the nonstop flow of digital cues. As a result, distraction becomes the norm, making it harder to be present with others—even when you're in the same room. Over time, this chronic distraction impairs the ability to experience deep connection. People are left feeling lonelier not in spite of, but because of, their connectivity.

As every moment is filled with digital signals, solitude—once a space to reflect and recharge—starts to feel unfamiliar and uncomfortable. People may experience restless dissatisfaction when alone (and even when they are with other people), compulsively checking digital devices and physical cues from people in order to feel included or validated.

The Universal Ache: When Everyone Is Lonely

Across decades of research and work with clients, one thread has become strikingly clear: The hunger for genuine connection runs deeper than age, gender, status, or achievement. Adolescents, mid-career professionals, retirees, entrepreneurs, caretakers, and others have sat in the same chair and described the same struggle to feel connected in the hyperconnectivity of today's world. Loneliness is not a niche experience or personal failing; it's a widespread reality, amplified by the ways dopamine-driven technology continually promises satisfaction but rarely delivers it.

What's most compelling is how digital platforms and constant stimulation have obscured—but not erased—the human need to be known, valued, and emotionally supported. Clients from every walk of life describe scrolling through highlight reels, messaging acquaintances, and swiping between group chats, all in pursuit of brief dopamine highs. Yet even as technology offers so many chances for virtual "touch," the rewards are fleeting; real intimacy and a deeper sense of trust between people are missing. Clients describe the experience of longing for connection amid the hyperconnectivity of social media as a "starvation in the middle of a feast." They are surrounded (through digital devices) by faces and noise, with no one truly feeling close to them. Men in executive suits, single mothers, college athletes, and retired couples have all spoken of lying awake, yearning for touch, laughter, and vulnerability that doesn't come through a screen.

The pain of loneliness is ancient, but the dopamine-fueled chase to ease that pain in modern times just makes it feel more potent. For some, even admitting to loneliness brings anxiety and self-doubt, as if struggle signals they are "failing" or "broken." However, when you acknowledge this longing, something transformative begins: You realize that this suffering is not weakness but a natural signal from the social brain. Dopamine drives you to seek connection, even when you don't find the reward you were hoping for. Honoring this universal ache—this shared human need—is the first step toward building connections that heal, protect, and satisfy in ways that technology alone cannot.

Rediscovering Connection: Science-Backed Practices for Renewal

Despite the powerful pull of digital distractions and dopamine-driven reward loops, loneliness is not something a person needs to struggle with forever. Modern neuroscience shows that authentic relationships built on vulnerability, attentive presence, and shared purpose leave lasting marks not just on the heart, but on the brain itself. Those moments of deep engagement activate dopamine and oxytocin together: a blend that delivers true emotional connection rather than fleeting pleasure.

Breaking the restless cycle of digital cravings involves practical strategies that retrain the reward system to light up via meaningful bonds rather than interactions on social media. Slowing down to enjoy device-free meals, engaging in conversation, and working together on collaborative projects nurture the brain's circuits of trust, joy, and connection—the very things that dopamine is meant to reinforce. Expanding your social world by meeting new people, listening to different perspectives, and joining purpose-driven communities builds both individual and collective resilience. It also boosts dopamine in ways that foster real belonging.

Practices in mindfulness and gratitude and spending intentional time alone also help rewire the brain from constant seeking to deeper contentment. Neuroscience confirms it: When you allow space for vulnerability and emotional honesty, dopamine is no longer just about anticipation or craving. Instead, it signals real reward, found in moments where you are truly present and feel known and emotionally safe. In addition to simple relief from loneliness, you find genuine, protective bonds that nourish your well-being.

In the next section of this chapter, you will take steps toward banishing loneliness and making true connections (steps you will continue in Part 3). As you move through this exercise, and the remainder of this book, remember that loneliness is an adaptive signal that was used to keep your ancestors safe. Rather than viewing it as shameful, try viewing it as a helpful prompt. Easing lonely feelings isn't about "fixing" yourself; it is about listening to and honoring your brain's ancient wisdom—the same wisdom that has preserved connection, community, and resilience for millennia.

CASE STUDY
Anna—The Invisible Wall

Anna's move to a new city marked the start of her career as a tech entrepreneur—her schedule filled with meetings, investor calls, and a rapidly expanding digital network. Unlike her upbringing in a tightly woven family, Anna's days were now saturated with brief professional exchanges and surface-level social interactions. At twenty-eight, her achievements were impressive, yet beneath the success, she felt an increasing sense of emptiness. The attention and praise she received online only sharpened this feeling of isolation.

Anna often found herself scrolling social media late into the night, searching for updates that would make her feel more involved, but nothing really landed. She sensed a disconnect, not only from close relationships back home but from herself—loneliness that buzzed beneath every digital notification. Where connection had always meant vulnerability and laughter in her family's kitchen, it now seemed replaced by performance and constant availability. Now she battled restless sleep, a decrease in motivation, and the silent question: Why is genuine intimacy harder to find as she grows more "successful"?

When Anna finally admitted her loneliness, she decided to show vulnerability. Letting go of her instinct to keep everything polished and professional, she opened up to a few trusted colleagues about her struggle. Some conversations felt awkward at first, but over time, they also became more real. She began carving out space for phone-free weekends, reaching out for deeper conversations, and reviving family traditions that provided comfort. With each step, she discovered relief from the exhausting chase for approval and uncovered the quiet power of just being present.

Anna's journey moved her from digital recognition to true belonging—a shift enabled not by more connections, but by her willingness to risk emotional honesty and rediscover genuine intimacy.

Practical Strategies: Authentic Belonging

Pause and practice: The most compelling insight from neuroscience is that what truly heals loneliness is not more *interaction*, but more *intentional and meaningful engagement*. Practical pathways to authentic belonging begin with clear choices: setting boundaries with technology, carving out protected time for actual conversation, and prioritizing relationships that invite honest connection instead of constant performance. You can follow these steps now to begin:

1. **Identify your moments of presence.** Start by identifying the moments each day when you can be fully present—at meals, on walks, during time spent with others—without the distraction of screens or notifications.

2. **Choose to invest in a handful of genuine, vulnerable relationships.** Let yourself be honest about needs and limits, even if it feels uncomfortable. Practice active listening, reflect back the feelings and meaning shared, and allow for silence to give space for deeper bonds to form. Dedicate at least two to three meaningful interactions weekly (or daily if possible) where you practice these deeper relational skills. Consistency matters more than duration; even fifteen to twenty minutes of undivided attention can strengthen neural pathways associated with secure attachment and emotional attunement.

3. **Embrace community.** Seek out opportunities for shared purpose— a community project, collaborative work, or creative pursuits with others—where the act of working together builds synchrony and mutual support. Aim for at least one collaborative or

community-oriented activity weekly, whether that's volunteering, group fitness, creative collaboration, or team-based work projects.

4. **Protect authenticity.** Notice when comparison or self-presentation starts to overshadow authentic exchange; gently redirect your attention to what is real, present, and mutual. This redirection will happen continuously within all interactions.

Why this matters: While these steps are neither prescriptive nor exhaustive, they are invitations to shift your attention from mere connection to belonging, from recognition to resonance. As the science and lived experiences show, lasting satisfaction comes not from gaining phone contacts or social media followers, but from meeting others—and yourself—where meaningful depth resides.

CHAPTER SUMMARY

In a world of constant connection, people are feeling more disconnected than ever. Fortunately, loneliness doesn't need to be a frequent struggle. The following are the key lessons to take from this chapter:

- Digital connection offers constant validation and social interaction without true fulfillment.
- Loneliness is a universal, evolutionary signal—a finely tuned alarm in your brain's reward system that seeks genuine connection and trust.
- The social brain relies on vulnerability, emotional honesty, and focused engagement to fully satisfy its dopamine-driven need for real bonds.
- Science-backed strategies for belonging, such as focused presence, mutual support, and purpose-driven relationships, are available to all, regardless of age or circumstance.

CHAPTER 7

Dopamine, Trauma, and Healing

Rewiring after Pain

Dopamine is the chemical of desire, learning, and survival. When trauma strikes, it does not simply leave an emotional wound; it rewires how the brain handles reward, motivation, and self-perception. Experiences of pain, neglect, or threat hijack the brain's dopamine signals, swapping out healthy anticipation and satisfaction for compulsive urges, numbing, withdrawal, and a deeper sense of emptiness. After trauma, many people find themselves in a relentless search for relief—sometimes through substances, achievements, relationships, or risky behaviors—only to find joy and pleasure remain out of reach.

This chapter digs deeper into how trauma alters dopamine pathways in the brain. The brain's capacity to adapt (known as **neuroplasticity**) is an invitation to transform pain into growth. Dopamine dysfunction can be rewired through a process that honors both the science and the suffering. In the following pages, you'll uncover how old survival strategies morph into modern compulsions, how the body and mind conspire to keep pain out of reach, and how even the smallest moments of connection, movement, or success can reignite the reward system. Recovery is not about erasing the past—it is about reclaiming agency, feeling, and authentic engagement with the world.

This chapter is for anyone who has felt the aftermath of trauma in their body, their habits, or their relationships. It is for those whose daily life has been colored by numbness, anxiety, or the persistent urge to escape. Here, you will learn the details of the trauma–dopamine connection, so that in Chapter 14, you will be ready to work more with trauma in guided practices. Drawing from real-world experience, you'll meet Eddie, whose childhood trauma and ultimate journey to healing offers a powerful testament to the brain's abilities. By combining neuroscience with lived experience, you can find new ways to awaken motivation, rediscover feeling, and build a foundation for lasting emotional healing. No matter how disrupted the dopamine system may be, repair is possible—joy and meaning can return to a life once defined by pain.

Trauma and Dopamine: How Adversity Rewires the Reward System

Trauma never leaves the brain untouched. Whether it is a specific event or chronic struggle, trauma sparks an adaptive response in the nervous system. Your brain circuits are flooded with stress hormones and the rhythm of **neurotransmitters**.

A CLOSER LOOK
Neurotransmitters

Neurotransmitters are chemical messengers that send signals across brain synapses. Each neurotransmitter has a distinct role, and when trauma disrupts the nervous system, it fundamentally alters how these chemicals are released, reabsorbed, and utilized.

Dopamine—the neurotransmitter that links anticipation, reward, motivation, and focus—is especially vulnerable to this trauma response. Repeated psychological stress or sudden emotional shock causes the brain's reward system to shift into survival mode. When the brain is in survival mode, reward becomes unpredictable; what once brought joy now barely registers, and habits that formerly fueled motivation can transform into rituals of relief or avoidance. The person may turn to compulsive behaviors (overworking, binge eating, substance use) that can temporarily mask the pain but only deepen the sense of numbness. At the same time, it becomes harder to make decisions and regulate emotions. Ordinary life feels exhausting or joyless.

Crucially, these changes are not signs of personal failure, weak character, or lack of willpower: They are the brain's effort to protect itself from further harm. Knowing this can bring comfort and clarity. The consequences of trauma are not "all in your head"—they are real, physical changes in the brain (neuroplasticity). This insight is the first step toward healing; in this knowledge, you can open the door for hope where pain had closed it. With this groundwork laid, you can dig further into how your emotions are impacted by trauma—and what to do about it.

A CLOSER LOOK
Neuroplasticity

As mentioned at the beginning of this chapter, neuroplasticity is the brain's ability to physically reorganize itself throughout life by forming new neural connections. When you have a traumatic experience, your brain wires together neural pathways that support survival (hypervigilance, fear responses, avoidance). These pathways become deeply grooved through repetition and emotional intensity, making the trauma response feel automatic and uncontrollable.

From Survival to Compulsion: The Emotional Fallout of Dopamine Disruption

As you discovered in the previous section, daily experience begins to shift in subtle and profound ways when trauma has impacted a person's brain chemistry. Most people associate trauma with sadness, fear, or anger, but the earliest and most persistent symptom is often **emotional numbness**—a dulling of feelings. This is the nervous system's way of protecting itself from further harm, minimizing the pleasure felt in order to also minimize the risk of getting hurt. A domino effect then follows: It becomes harder to make decisions and adapt to changes; focus drifts, and ordinary moments feel flat or disconnected.

A CLOSER LOOK
Emotional Numbness

Emotional numbness is not indifference; it's a protective mechanism. When dopamine signaling flattens in response to chronic threat, your brain essentially dampens the reward and pleasure systems. Simultaneously, the **amygdala** (your threat detection center) becomes hyperactive, while the prefrontal cortex (responsible for emotional regulation and assigning meaning to events) shows reduced activity. The result: You can perceive events intellectually but struggle to feel their emotional weight. You recognize that something should matter, but the signal, the dopamine-driven sense of significance, fails to register.

Compulsions emerge over time because the dopamine system, which once supported healthy motivation, starts seeking relief from discomfort. The *drive* to feel pleasure grows, even as the *ability* to feel it fades. In these situations, some people turn to substances, excessive work, risk-taking, or repetitive rituals, hoping for a sense of satisfaction that never fully arrives. These behaviors are not simply "bad choices," but the brain's desperate attempt to fill a neurochemical void.

As familiar coping mechanisms stop working, the emotional burden deepens. Anxiety rises as the prefrontal cortex struggles to regulate impulses and keep danger in check. Depression may also set in, not just as general sadness but as an absence: the inability to feel, imagine, or hope. Over time, relationships often suffer, strained by irritability, withdrawal, or a lack of connection. Trauma warps how every experience is perceived. For some, it leaves them on constant alert, while others shut down or, most often, get caught in ongoing stress.

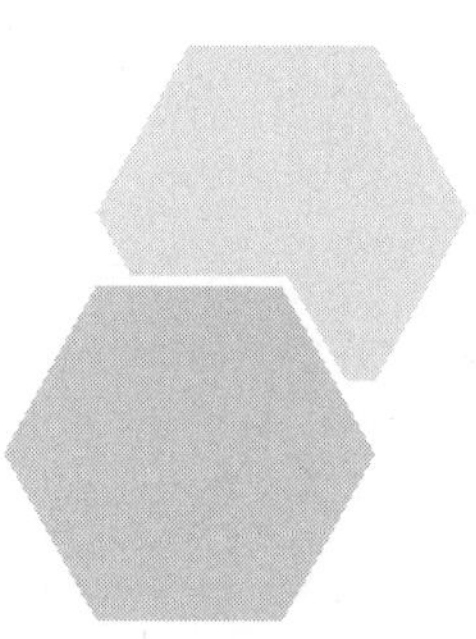

The consequences of trauma are not "all in your head"—they are real, physical changes in the brain (neuroplasticity).

This insight is the first step toward healing; in this knowledge, you can open the door for hope where pain had closed it.

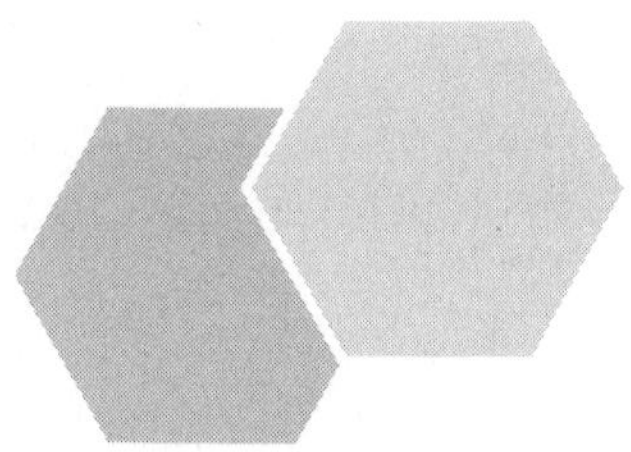

The Science of Healing: Rewiring Dopamine after Trauma

Recognizing compulsive behaviors, emotional numbness, and stress responses as the brain's way of surviving trauma sets the stage for healing. These patterns are not character flaws, but mechanisms that the brain can unlearn and rewire through focused, science-based treatment. Modern neuroscience reveals that healing is a deliberate process rooted in the brain's ability to adapt and restore balance in its dopamine pathways. When adversity blunts or warps these circuits, modern interventions are designed to reignite hope, motivation, and real emotional engagement at the molecular level. The remainder of this section explores methods for healing on this deeper level. Rather than approaching these methods as a collection of isolated techniques, each can work in tandem with others to restore your capacity for hope, motivation, and meaningful engagement with life.

Innovative therapies, such as **neurofeedback**, help clients to retrain their brains for improved emotional balance and self-control. By learning to regulate specific neural circuits, they are able to use neurofeedback to restore clarity, focus, and sensitivity to reward. In both laboratory and clinical settings, this approach has revolutionized healing for trauma survivors. It can restore true joy and motivation without dependence on medication alone.

Other behavioral therapies, like **cognitive remediation therapy**, can bring about sustainable change. Cognitive remediation therapy addresses specific deficits in attention, memory, and **executive functioning**. This therapy is used to rebuild neural pathways that have been disrupted by trauma and dopamine dysfunction. Trauma-informed mindfulness programs such as cognitive remediation therapy are deeply grounded in brain science and help people regulate their emotions, reduce stress, and process difficult memories, which in turn fuels new and resilient dopamine responses.

A CLOSER LOOK
Neurofeedback

Neurofeedback is a brain-training technique that uses real-time monitoring of brain activity to help you develop greater control over neural functioning. During a neurofeedback session, electrodes are placed on your scalp to measure electrical activity in specific brain regions. This data is converted into immediate visual or auditory feedback, typically displayed on a screen. Your brain learns to recognize and strengthen patterns of activity associated with calm, focus, and emotional regulation. For trauma survivors, neurofeedback is particularly effective because it targets the dysregulated brain circuits directly, bypassing the need to verbally process trauma while still achieving lasting rewiring of dopamine and threat-detection pathways.

Another tool is **personalized restoration of dopamine function**. Here, clinicians use medications that act directly on dopamine receptors and transporters. These targeted treatments enhance motivation, executive function, and emotional flexibility. Paired with cognitive training and behavioral therapies, personalized dopamine restoration accelerates healing where standard antidepressants may fall short.

A CLOSER LOOK
Executive Functioning

Executive functioning encompasses the cognitive processes that allow you to plan ahead, make deliberate choices, regulate impulses, shift your attention flexibly between tasks, and organize complex information. It's the force behind purposeful action.

Perhaps most transformative to trauma healing are the ongoing advances in **precision brain health**. These approaches focus on genetics and **biomarkers**, allowing for individualized plans that rebuild healthy dopamine balance, repair reward systems, and guide trauma survivors from emotional numbness into authentic recovery.

⌕ A CLOSER LOOK
Biomarkers

Biomarkers are measurable biological indicators that signal the presence or progression of a physiological state. In dopamine and trauma research, biomarkers might include levels of stress hormones or brain data showing activity patterns in the reward circuitry. These objective details help researchers track how trauma alters brain chemistry and monitor recovery progress over time.

The message from the latest research is clear: No matter how severe trauma's impact, the human brain has the power to repair itself and flourish. Every dopamine circuit can be strengthened or rewired, laying the foundation for genuine joy.

A Daily Practice: Routines for Dopamine Renewal

Scientific advancements are only as impactful as their translation to daily life. Before more targeted treatment and therapies for traumatic stress can be considered, it is important to create (and follow) a structured routine: Regular sleep, nutrition, and physical movement establish the foundation for balanced dopamine and trauma recovery. Consider including consistent exposure to natural light, physical activity, and rhythmic movement such as brisk walking, swimming, or group exercise in this routine. These simple experiences stimulate the growth of dopamine receptors and enable a deeper emotional reawakening following trauma.

Cognitive remediation practices, previously reserved for clinical settings, can also now be incorporated into these routines. These practices include targeted brain exercises that improve attention, working memory, and executive functions—skills often dampened by trauma.

By intentionally challenging the brain through puzzles, learning new skills, or engaging in mindful problem-solving, you can open new pathways for dopamine release and activate the brain regions that are critical for focus and decision-making.

The practice of trauma-informed mindfulness is another cornerstone of a healing routine. Guided awareness exercises, breath work, and focus on the present moment help regulate stress responses, reduce emotional reactivity, and reestablish a healthy anticipation for rewards. Daily mindfulness not only addresses anxiety and numbness, but also cultivates pleasure, meaning, and connection—bridging gaps left by chronic dopamine dysfunction.

Once you have a supportive routine in place, you can add more specific treatments such as the methods explored earlier. Therapeutic relationships are arguably the most potent ingredient in recovery. Intentional work with a trusted mentor, coach, or therapist provides both safety and novelty—key triggers for dopamine renewal. By creating opportunities for authentic connection, emotional risk-taking, and meaningful reflection, these partnerships become the brain's foundation for sustained change.

Recovery from trauma is neither a solo act nor linear; it is the cumulative result of science, lived experience, and a compassionate support network all working together.

Agency—The Heart of Dopamine Recovery

Amid the healthy routines and science-based methods, it's important to keep in mind that the core of rewiring dopamine pathways after trauma is agency. The belief and ability to direct your life with intention and hope is what leads to success in healing. Trauma often strips a person's sense of agency away, making motivation difficult to find and daily decisions a struggle. Neuroscience shows that agency is not just a feeling, but also a physical reality of the brain: Repeated intentional actions strengthen neural circuits and boost the release of dopamine.

The act of reclaiming agency is as much biological as it is experiential: The dopamine system thrives on autonomy, choice, and the pursuit of meaningful goals. By harnessing their own power, survivors can rewrite the narrative of trauma.

Taking agency begins with small, consistent choices. Whether it's setting boundaries, initiating new relationships, or exploring previously neglected interests, each act of personal advocacy retrains the brain to anticipate and experience reward. Empowered choice also helps transform compulsive patterns. When people are faced with urges to repeat old habits, even a moment of reflection or redirection—such as pausing before acting or substituting a healthy alternative—can interrupt those temptations and spark change. Over time, these new responses accumulate, supporting healthier behaviors and gradually restoring feelings of pleasure, confidence, and self-worth.

Resilience—Another Cornerstone of Healing

Cultivated alongside agency, resilience is another important part of rewiring dopamine after trauma. Resilience grows from how you engage with setbacks. Seeing challenges as opportunities to learn, rather than evidence of failure, strengthens the brain's reward system and supports recovery. This act of self-compassion paired with realistic optimism helps maintain progress amid the inevitable ups and downs.

Sustainable healing from trauma is not a single breakthrough—it is a collection of persistent efforts, adaptive practices, and deep personal growth. Resilience is forged across *repeated* moments where not just setbacks, but also discomfort and doubt, challenge progress. It's easy to think that enduring these struggles wears on the brain, but neuroscience shows that each recovery effort—no matter how small—stimulates neural growth, strengthens reward pathways, and builds adaptability.

Central to resilience is the practice of pattern disruption. Survivors can cultivate resilience by intentionally stepping away from old habits and embracing new, healthy routines. This might mean adding mindfulness exercises to your morning rituals, choosing nourishing social connections such as vulnerable conversations, or prioritizing sleep. Routine, novelty, and self-care all send powerful signals to the brain that boost dopamine and reinforce emotional strength.

As resilience grows, emotional regulation becomes stronger. Stressful memories or **trauma triggers** lose their power when the brain is trained to observe, reflect, and redirect (rather than react automatically). By practicing techniques such as deep breathing, gratitude journaling, or engaging in creative pursuits, you can actively reinforce neural circuits that support calm, perspective, and hope.

A CLOSER LOOK
Trauma Triggers

A trauma trigger is any internal or external cue that activates the memory of past trauma and sends your nervous system into a stress response, even when no actual threat is present. Triggers can be sensory (a sound, smell, or visual image that resembles aspects of the traumatic event), contextual (a specific location or time of day), relational (an interaction pattern that echoes the original trauma), or emotional (a feeling state that was present during the trauma). In response to a trauma trigger, you may find yourself flooded with panic, rage, or dissociation.

Resilience also depends on supportive relationships, guided self-reflection, and a willingness to adapt. Working with therapists, mentors, or trusted peers—people with different perspectives who can hold you accountable for your goals—are key to keeping your momentum. With every intentional act, the brain rewires itself for resilience, revealing that lasting change is both a science and an art, available to anyone willing to embark on the journey.

CASE STUDY
Eddie—Rewiring a Life after Trauma

Eddie grew up in a world shaped by uncertainty—a working-class neighborhood, an alcoholic parent, and evenings spent tiptoeing around conflict and neglect. For years, Eddie's brain learned to survive by numbing his emotions, avoiding risk, and seeking small comforts in routine. As he grew into an adult, the hurt of his trauma became harder for him to escape. He worked long hours as a school janitor, lost himself in repetitive tasks, and felt both invisible *and* indispensable. Daily life was a cycle of exhaustion, anxiety, and a haunting sense of emptiness that no midday snack or television binge could fill.

The cost of trauma showed up in subtle ways. Eddie struggled to feel pleasure: The laughter of children in the hallway felt distant, and even a pat on the back from a colleague failed to spark joy. Nights were spent wrestling with compulsive thoughts—checking door locks, reviewing to-do lists, and worrying about loved ones. Attempts to break these patterns only led to frustration. Motivation came in short, unpredictable bursts, quickly replaced by periods of numbness or anxiety.

When Eddie first came to me, he described how years of pushing through pain left him feeling disconnected from life. Our work together began with a careful assessment of his emotional responses, habits, and thinking patterns. I structured our sessions around neuroscience-based approaches: building a foundation of restorative routines, cognitive remediation exercises for attention and focus, and trauma-informed mindfulness to develop emotional regulation. We emphasized gradual progress, celebrating small victories and pausing thoughtfully during setbacks.

Over time, Eddie noticed fundamental changes. Moments of laughter felt sincere, his relationships deepened, and the exhausting pull of compulsions gradually faded. Eddie's journey was not linear; setbacks challenged his progress. But each effort was supported by compassion, professional guidance, and an understanding of how the brain adapts. The rewards were undeniable: Eddie experienced a renewed sense of agency, an ability to find meaning in ordinary moments, and emotional resilience where there had once been only pain.

Eddie's story reveals that healing is possible, no matter how long trauma has shaped a person's life. Rewiring the dopamine system can restore not only motivation, but also the genuine pleasure and depth that define a life well lived.

Practical Strategies: Rewiring Trauma for Dopamine Renewal

Pause and practice: The following are carefully crafted neuroscience-based strategies I have used with clients to help reframe trauma, rebuild dopamine balance, and reclaim emotional vitality. Drawn from the ideas explored in this chapter, these proven approaches can be used to take control in the wake of trauma:

1. **Audit and interrupt old patterns.** Begin by keeping a daily log of emotional triggers, automatic thoughts, and compulsive habits. Set intentional times to pause and check in with your log (midday breath work, short reflective breaks, or digital-free moments allow you to notice and disrupt compulsive responses shaped by trauma). This creates a feedback loop: You record what you observe and then review those entries during your intentional pause points to identify recurring patterns.

2. **Reframe your narrative through neuroplasticity.** After establishing your tracking practice for 2–4 weeks, begin the deeper work of changing the story in your mind. Dedicate time each day to consciously reframe memories and pain surrounding trauma. Use journaling or spoken reflection to reconstruct the story you tell yourself about the past, focusing on points of resilience, growth, and possibility. This activates and strengthens new brain pathways for agency and hope.

3. **Swap avoidance for meaningful engagement.** As you establish your tracking and reframing practice, start experimenting with restorative alternatives. Instead of retreating into habitual avoidance, actively choose restorative alternatives such as brief exercise, learning something new, expressing yourself creatively, or connecting with a trusted friend. Each healthy action is a neuroplastic boost of dopamine, triggering the release of this motivation neurotransmitter and making pleasure and forward momentum more accessible over time.

4. **Create sanctuary spaces and rituals.** At the same time, designate calming environments—such as a favorite chair, a walking route, or a quiet corner—as safe zones to check in and regulate your emotions. Practice restorative routines such as meditation, deep breathing, or mindful reflection to reinforce positive neural circuits and build resilience.

5. **Prioritize authentic connection.** Alongside the previous exercises, schedule regular moments for genuine social engagement, whether through in-person interactions, group activities, or heartfelt conversations. Social neuroscience confirms that compassionate relationships supercharge dopamine and counteract isolation—fueling meaningful change.

6. **Use accountability and guided support.** This is best introduced once you have established some consistency with the foundational practices mentioned previously (typically 3–4 weeks in). Share your intentions and progress with a trusted mentor, therapist, or supportive peer group. Use the social reinforcement of regular check-ins and collective celebrations of wins to keep you motivated and moving forward.

Why this matters: By consistently practicing these strategies, you'll reshape old, trauma-based routines, boost focus and motivation, and use neuroscience to track and celebrate your progress—one intentional day at a time. In Chapter 14, you will find more practical steps for working with dopamine to ease the effects of trauma.

CHAPTER SUMMARY

Trauma, whether a single event or prolonged harm, fundamentally alters the dopamine systems in your brain. Here are the main things to remember about dopamine and trauma from this chapter:

- Trauma causes the brain's reward system to shift into survival mode, where joy is diminished and unhealthy coping mechanisms develop.
- Avoidance, withdrawal, and the use of substances are common coping mechanisms for trauma that can offer temporary relief. However, they do not lead to true healing.
- Reframing the story trauma has created in your head, replacing old habits with healthy and healing routines, and prioritizing real-world connections actively reshape the brain's reward system.
- Advances in neuroscience also provide targeted methods for healing and balancing dopamine after trauma, including neurofeedback therapy and personalized dopamine restoration.

The Dopamine Comeback

Rediscovering Joy, Drive, and Lasting Satisfaction

Now that you have learned about the dopamine system and how things like advancing technology and trauma can disrupt its natural flow, it's time to dismantle these obstacles and find authentic motivation, joy, and well-being. In Parts 1 and 2, you cultivated an understanding of where feelings like pleasure and anticipation come from, and you tried practical strategies for reflecting on your own dopamine pathways and navigating challenges such as social media. In Part 3, you will continue this work in more depth. This part helps you turn insight into action, translating brain science and comprehensive tips into personalized practices that energize your daily life. If you've ever felt that happiness and fulfillment feel short-lived in a world of endless stimulation and competing demands, this part offers not just hope, but a step-by-step blueprint for sustainable change.

Across seven chapters, you'll learn to create your own "Dopamine Menu"—a tool kit of routines and strategies that matches your unique strengths, values, and lifestyle. You'll see how dopamine detoxes can temporarily reset your reward system, but also why real change relies on steady and intentional effort. Each chapter builds on the last, showing you how to use motivating environments and habits to automate progress and how movement, nutrition, and sleep fortify the brain's chemistry for focus and drive.

At work, you'll find targeted, brain-based methods for cultivating concentration, managing energy, and staying satisfied—even when life gets stressful or complex. You'll also tap into creativity through novelty and healthy risk-taking, using the same dopamine pathways that drive motivation—without burning out. Most importantly, every chapter helps you build resilience and restore dopamine balance through personalized recovery rituals and practical planning.

Throughout, neuroscience and real-world experience combine to give you clear, actionable steps toward satisfaction, momentum, and well-being. You'll gain the tools to review (or "audit") your routines, redesign your days, and create rhythms that protect your energy—whether you're working, creating, caring, or healing. By the end of this part, you will be equipped not just with knowledge but real techniques for finding lasting joy, motivation, and self-driven fulfillment. This is where your comeback begins—an invitation to rediscover purpose, drive, and the full spectrum of well-being that a healthy dopamine system can offer.

CHAPTER 8

The Dopamine Menu

Building Your Personalized Tool Kit for Joy and Focus

Enduring motivation and emotional well-being aren't built on short bursts of pleasure. They're cultivated through discerning which daily rituals and experiences truly nourish your brain's reward system. As you uncovered in Part 1, dopamine's role is more nuanced than simple gratification; it sparks focus, anticipation, creative flow, and the drive to pursue meaningful goals. Crucially, the conditions that trigger, sustain, and replenish dopamine are as unique as your own emotional history, genetic wiring, and lived environment.

In this chapter, you'll learn more about the "Dopamine Menu." Rather than relying solely on prescriptive routines or one-size-fits-all advice, the concept of a Dopamine Menu means you get to create an array of activities, habits, and rituals that directly match your brain's needs. Think of it as building a personalized workout routine for your brain—a mix of productivity sprints, restorative breaks, challenging pursuits, and connection rituals that offer satisfaction and energy in authentic, sustainable ways. The menu can also be adjusted, expanded, and refined based on how your brain responds to changing seasons, different stressors, novelty, and recovery.

Current research suggests a key insight: Intentionally building and regularly revising a personalized Dopamine Menu makes it possible to avoid burnout, monotony, and emotional volatility. Instead, you can create reliable pleasure, focus, and resilience. Learning from Jack, you will also see a real-world example of how a curated menu can reignite a spark that has been dulled. With a well-designed Dopamine Menu, you move beyond superficial hacks and cultivate a daily practice that transforms routine into reward and effort into fulfillment.

Your Personalized Dopamine Menu: A Blueprint

Building a Dopamine Menu is both an art and a science: It takes observation, experimentation, and a willingness to honor your unique brain. The goal is not to follow a universal recipe but to assemble an ever-evolving selection of options that foster sustainable drive, pleasure, and emotional stability.

To begin creating your own menu, you'll need to start with a "menu audit": Explore which routines, activities, and social interactions most reliably make you feel engaged, motivated, and calm. If you are unsure, you can use daily logs or digital trackers to note when you get a healthy dopamine boost. When does anticipation spike? Which moments deliver true satisfaction? Which moments come with only brief pleasure or a post-activity crash? Pay close attention to how your energy, focus, and mood fluctuates each day as you work, interact, and rest.

Establishing the Menu Courses

Once you have identified the things that deliver true satisfaction, categorize these menu items by function:

- **Main courses:** Activities that deliver deep engagement, such as focused projects, immersive learning, or creative expression, are the staples that regularly recharge your reward system. Main course items are closely aligned with the "creative challenges" mentioned later in this chapter: Both are activities that engage the prefrontal cortex, require sustained focus, involve problem-solving or skill development, and deliver intrinsic satisfaction through progress and mastery. Examples of main dishes include learning a new software platform, composing music, and mentoring or teaching others.

- **Side dishes:** These are short bursts of novelty, movement, or brief social connection that complement your main courses and keep you from feeling mentally fatigued. Side dishes serve as quick dopamine resets between periods of intense focus, preventing cognitive fatigue and mental sluggishness. They can include a ten-minute walk, a quick phone call with a friend or colleague, and a few minutes of listening to music or watching a short video. Side dishes typically last five to fifteen minutes.
- **Restorative desserts:** These are the essential ingredients for mental and emotional renewal, including mindful relaxation, nature walks, music, meditation, and sleep.

Experiment with the sequence and timing of each course to optimize the appeal of every item each time you enjoy it. In addition, rotate new "specials" into each course. These infrequent items—trying unfamiliar hobbies, shifting your work rhythms, exploring new relationships or environments—will help personalize your menu.

Experimenting, Tracking, and Adapting Your Menu Items

A Dopamine Menu that delivers lasting fulfillment involves more than just selecting the right items. It requires tracking and adjusting each menu item over time and experience. The Dopamine Menu is a living system, where every piece evolves with the flow of life. Approach each menu item as an experiment in self-discovery. Daily selections—whether it's early morning writing sessions, afternoon walks, brief bursts of playful networking, or periods of concentrated learning—should be treated as working hypotheses that invite investigation and refinement. Each "dish" can be sampled, adapted, and even retired based on the feedback you get from your mind and body.

Working with your living menu involves being meticulous with tracking. Use digital apps, mood diaries, or even color-coded calendar notes to mark changes in drive, inspiration, and emotional stability

The Dopamine Menu is a living system, where every piece evolves with the flow of life. Approach each menu item as an experiment in self-discovery.

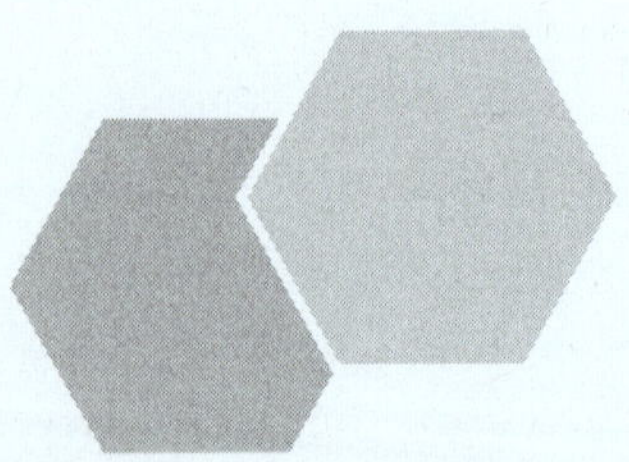

across days and weeks. Instead of focusing only on peak moments, reflect on the tone and aftereffects that linger in more subtle moments as well. Does a creative sprint (such as focused writing or design) leave you re-energized or slightly depleted? Does a long lunch with a colleague add a sense of meaning, or is it only a fleeting distraction? Over time, these observations paint a kind of neurochemical "map"—a personalized chart of what lifts your mood, inspires clarity, and builds satisfaction.

Use broader reviews (biweekly or monthly menu check-ins) to identify greater patterns, celebrate menu choices that are effective, and gently eliminate those dishes that no longer enhance your life.

Additionally, adapt your course proportions based on your evolving needs: Intense periods of life may require extra restorative rituals (desserts), while periods of strong momentum and inspiration can call for supplementary challenges (main courses). Some people even run themed "menu challenges," varying one element each week, such as the timing of restorative breaks or the type of connection ritual. Observe the impact; if an activity or ritual ceases to nourish you—meaning, it leads to emotional flatness, impatience, or restlessness—you can replace or rework it.

Sometimes more profound transformations arise from intentionally simplifying the menu: Try removing excess stimulation, reducing multitasking, or introducing slow-paced, contemplative time. It is also important to honor the role of emotional turbulence and setback. Not every week will produce breakthroughs; plateaus and steps backward are signs your menu is ready for an upgrade.

Making Compassionate Changes to the Menu

As you track and adjust your Dopamine Menu, it's essential to avoid unhelpful guilt or blame. Instead, approach it with self-compassion, reframing every experiment and change as a stepping stone. Through consistent review, open-ended exploration, and acceptance of change, you build a Dopamine Menu that is both stable and responsive—a menu that turns ordinary rituals into powerful engines of engagement, resilience, and joy. The result is a life anchored in self-awareness, where each update brings you closer to authentic motivation and lasting well-being.

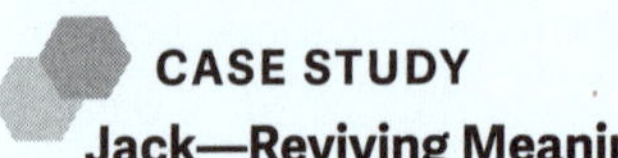

CASE STUDY
Jack—Reviving Meaning in the Ordinary

Jack was a skilled data analyst known for his accuracy, reliability, and expertise. Yet, after several years in this role, his daily routine began to feel repetitive; he found himself completing complex tasks easily, but with increasing emotional detachment and less satisfaction. Jack wasn't burned out—he was bored. And his dopamine system reflected this through an absence of anticipation, dwindling creativity, and pervasive restlessness.

As Jack and I began working together, his initial "menu audit" revealed that he was overly relying on repetitive main course activities— analyzing spreadsheets, reviewing and correcting data, and participating in structured meetings. Rarely did he include restorative desserts or novel side dishes. While his work was essential, a lack of variety, personal challenge, and rewarding closure made it unsatisfying.

Together, we crafted a Dopamine Menu with far greater diversity and intention. I encouraged Jack to introduce menu "specials": collaborations with other departments at his company, short creative assignments, skill-building sessions in data visualization, and targeted brainstorming meetings to break up the routine cycles. Restorative dessert options included taking scheduled walks, listening to music, and connecting with professional networks for fresh perspectives.

As Jack refined his dopamine menu, his days started to feel more engaging and satisfying. Weekly check-ins of the menu helped him keep novelty alive, match new duties with evolving skills, and remain emotionally invested, even when schedules became busy. His relationship to his work changed: Instead of waiting for external excitement, Jack found reward in shaping his own cycles and savoring subtle victories.

Jack's story demonstrates that motivation doesn't always require dramatic change; sometimes, it's about cultivating new flavors, rhythms, and ingredients in the everyday. A well-designed Dopamine Menu can reinvigorate even the most structured routines, restoring meaning, drive, and emotional balance.

Advanced Dopamine Menu Building: Adapting for Life Stages and Challenges

Alongside day-to-day changes, a Dopamine Menu is most effective when it is adapted for different seasons of life, changing demands, and individual brain needs. While the foundational elements (experimentation, reflection, and adaptation) remain central, the menu's "dishes" should evolve in tandem with your life circumstances, goals, and emotional landscapes.

For those entering new phases of life, such as early adulthood, their Dopamine Menu may lean heavily on novelty-driven pursuits (side dishes) such as career exploration, forming new social groups, and acquiring new skills. Midlife often calls for integrating more restorative practices (desserts) to counterbalance the rising stress and fluctuations of identity.

During major life transitions—such as career changes, parenthood, or relocation—it helps to do a menu audit and introduce new "dishes" that are relevant to shifting energy, needs, and priorities. Incorporating feedback from trusted partners, mentors, or social circles can also add fresh perspectives and prevent menus from growing stale.

Ultimately, advanced menu building is about honoring your brain's highly specific and evolving needs while staying open to more change down the line. No single menu template fits every stage or challenge, but by regularly customizing and experimenting with your Dopamine Menu, you can invite motivation, clarity, and joy throughout every phase of life.

Neurodivergence, Mood Instability, and Dopamine Dysregulation

People experiencing chronic distraction, mood instability, or **neurodivergent** profiles (like ADHD or high sensitivity) require a Dopamine Menu that maximizes variety without overwhelming the senses. Here, specialized "appetizers" (quick and low-effort practices) may be required to interrupt impulsive habit loops, while structured main courses and inventive "specials" (intentional rewards) can help sustain engagement, build routine, and foster resilience. Examples of dopamine appetizers include a two-minute breathing practice that resets your nervous system, responding to a single email or completing one small admin task (accomplishment dopamine), selecting music or a three-minute sensory moment, a brief text to a friend, or a quick body scan. Helpful main courses here include daily meditation, scheduled sleep windows, regular walks, predictable social touchpoints (such as a weekly dinner with close friends or a standing lunch date), or routine creative sessions. Specials may be project milestones, published work, completed challenges, an award, or a congratulatory gathering with family.

A CLOSER LOOK
Neurodivergence

Neurodivergence refers to variation in how brains are wired and how they process information. Neurodivergent individuals have brains that function differently from those of neurotypical individuals (those with conventionally wired brains). Standard neurodivergent profiles include ADHD, autism spectrum disorder, dyslexia, and sensory processing sensitivity. Neurodivergent brains often exhibit distinct patterns of reward processing, attention regulation, and dopamine sensitivity. For example, individuals with ADHD may experience dopamine peaks from novelty and stimulation. Those on the autism spectrum may find intense focus rewarding while experiencing sensory overstimulation from unpredictable social demands.

For those dealing with a dysfunctional dopamine system (such as emotional numbness or depleted motivation), menus should include both specials and low-pressure main course items like nature observation, unhurried creative exploration, or meaningful conversation. This combination will help reintroduce pleasure and anticipation at a manageable pace.

Acute and Chronic Sleep Issues

Sleep is another big factor when building your unique Dopamine Menu. When you sleep well (seven to nine hours for most adults), your dopamine receptors remain responsive and motivation feels accessible. Sleep deprivation, however, can at first provide a short-term dopamine spike (the wired, restless feeling), while chronic sleep loss blunts dopamine receptor sensitivity over time. The result is flattened motivation, less pleasure from usually satisfying activities, and increased cravings for intense stimulation. If you find your current Dopamine Menu ineffective, examine your sleep patterns and adjust if needed before making any menu changes. A week of improved sleep often restores menu items to their usual effect.

Also, work with your natural sleep patterns by scheduling main course activities during dopamine peaks after waking (typically morning or early afternoon for some). Reserve desserts and lighter side dishes for late afternoon and evening when dopamine naturally wanes before bed.

Hormone Cycles

Hormone cycles are another important element to factor into the Dopamine Menu, particularly for menstruating individuals. During the **follicular phase** of the menstrual cycle (the first half of the cycle that begins after menstruation ends and spans roughly two weeks), when dopamine sensitivity increases, main course activities may feel more accessible and engaging. However, the stress hormone, **cortisol**, is usually elevated at this time: Individuals can balance this boost with deliberate restorative dessert items to prevent overstimulation or burnout.

🔍 A CLOSER LOOK
Cortisol

Cortisol is a steroid hormone your adrenal glands release in response to stress and your **circadian rhythm** (your body's internal twenty-four-hour clock, regulating biological functions such as sleep-wake cycles, hormone release, body temperature, appetite, and dopamine signaling). In the right amounts at the right times, cortisol wakes you up, sharpens your focus, and gives you the energy to act. It peaks about twenty to forty minutes after you wake up, then gradually declines throughout the day so you can wind down at night. When your circadian rhythm is disrupted or stress is constant, cortisol becomes dysregulated, staying elevated when it should drop or remaining flat when it should rise. Chronically elevated cortisol impairs memory, weakens immune function, and dampens dopamine signaling, all of which erode motivation and make it harder to focus or engage with anything meaningful.

During the **luteal phase** of the menstrual cycle (the second half of the cycle, spanning roughly two weeks before menstruation, when progesterone rises and dopamine drops), the same main course item that felt energizing two weeks prior may now feel less rewarding or accessible. To adapt their Dopamine Menu during this phase, individuals can shift toward more restorative desserts, such as a warm bath, gentle stretching, or time in nature without performance pressure. They can also emphasize side dish items that feel lighter and require less sustained focus, such as journaling, listening to music, or organizing a single drawer, and extend recovery time after intensive work.

Beyond the menstrual cycle, those experiencing perimenopause or menopause should emphasize community connection, movement-based side dishes, and restorative "desserts." Similarly, hormonal shifts during pregnancy, postpartum recovery, and andropause (in men) can make certain menu items feel less effective. Rather than abandoning the Dopamine Menu entirely, recognize these as seasons requiring some adjustments.

Navigating Pitfalls and Roadblocks: Staying On-Menu

Even the most thoughtfully designed Dopamine Menu will encounter obstacles from time to time. These may include old habits that refuse to budge, periods of emotional flatness, or a feeling of overwhelm from too much novelty. Neuroscience shows that these setbacks are part of adaptation and survival; the brain's reward pathways can easily be disrupted by stress, monotony, or competing demands.

Common pitfalls to staying on-menu include mistaking quick bursts of stimulation—such as impulsive screen time, relentless multitasking, or constant novelty—for genuine satisfaction. Relying too much on dessert menu items, such as easy comfort or distraction, can drain motivation and leave your dopamine system depleted. Likewise, cycles of overcommitment to high-intensity main courses may come with temporary drive but result in exhaustion or burnout if no restorative dessert items are included to balance them out.

To stay the course, expect plateaus—moments when menu favorites lose their spark or stop feeling rewarding. Treat these moments as helpful feedback for your menu. They are an invitation to adjust course proportions and add or subtract items. Celebrate small wins—breakthroughs in energy or awareness, periods of restored clarity, or even the honest recognition that you need rest or a change.

Prepare for life's unexpected shifts: stress, grief, relationship conflict, or sudden opportunity. These larger changes can disrupt even the most balanced menu, but staying flexible and keeping this chapter at the ready will help you adapt. Practice self-compassion; every setback is data for refining the next iteration, not an indictment of personal failure. In addition, seek support from friends, mentors, or professional guides. They can help you notice unhelpful patterns and suggest new menu selections. Group check-ins, accountability partners, or collaborative menu planning can keep you motivated and emotionally steady during challenges.

With time, navigating these roadblocks and self-reflecting will lead to a menu that's more robust, personalized, and sustainable—ensuring balanced dopamine and well-being through every season of life.

 **CASE STUDY**
Naomi—From Chaos to Creative Flow

For Naomi, a creative director, her work life was filled with last-minute deadlines and constant novelty. The relentless churn of urgent projects lit up her dopamine pathways, but left her feeling exhausted, emotionally scattered, and unsatisfied. Her dopamine system was hooked on intensity—every rush of anticipation was followed by fatigue and a sense of emptiness.

When she arrived for coaching, Naomi's first "menu audit" revealed a heavy focus on thrill-seeking main course activities, with little space for restorative or celebratory desserts. Together, we mapped out how her work and emotional rhythms impacted her unique brain patterns, tracking which activities boosted drive and which led to burnout.

Instead of generic "work-life balance" advice, Naomi learned to create a Dopamine Menu that matched her unique cycles. She built in restorative desserts—quiet morning rituals, brief movement breaks, reflective pauses after major pushes—to smooth her transitions and anchor satisfaction. She experimented with alternative side dishes—collaborative brainstorming, learning new creative techniques, and spontaneous connection—that layered novelty without overwhelming her system.

Gradually, Naomi's menu shifted from frenetic urgency to purposeful flow. Her motivation became more consistent, her emotional highs weren't so intensely counterbalanced with lows, and her sense of fulfillment grew deeper and more reliable. Creativity flourished not in chaos, but in rhythms aligned with her brain's actual needs. Relationships strengthened as she made space for genuine connection rather than transactional exchanges. Naomi's success was not accidental—it was the result of a living, personalized Dopamine Menu designed and refined in step with her growth and circumstances.

Her journey demonstrates that the most powerful shift does not come from living a less intense life, but from curating and balancing it—transforming a roller coaster of deadlines into a creative menu that offers lasting clarity, mastery, and joy.

Practical Strategies: Building Your Personalized Dopamine Menu

Pause and practice: Now that you've dug deeper into the Dopamine Menu and tips for personalization, it's time to bring your own Dopamine Menu to life. Simply follow these science-driven steps:

1. **Conduct a menu audit.** For one week, record every activity, social exchange, and environmental shift that impacts your energy, engagement, and mood. Treat each entry as a menu item, making note of frequency, intensity, and aftereffects.

2. **Balance your courses.** Use your audit to create a menu with a clear variety of main courses (focused work, deep creative flow, or meaningful conversations), side dishes (novel skills, positive micro-moments, or quick collaborations), and desserts (restorative rituals, nature, music, or truly rejuvenating solitude). Adjust course portions and frequency as needed, taking into account the different seasons of life.

3. **Rotate in specials.** Prevent habituation by introducing fresh "dishes"—a new project, hobby, learning opportunity, or social ritual—especially when boredom or emotional numbness begins to set in. If something loses its effect, substitute or modify until a sense of joy and active engagement returns.

4. **Time your consumption.** Learn your body's dopamine rhythms to optimize when menu selections are most satisfying. Schedule main course activities for when you have peak energy, and plan dessert items after intensive tasks to maximize reward and reinforce healthy cycles.

5. **Reflect and revise regularly.** Review your menu items at weekly or monthly intervals to identify what truly sustains your drive and fulfillment. Remove old staples that have lost their appeal, and embrace rituals that provide the most lasting value.

6. **Watch for overindulgence.** Monitor warning signs—restlessness, irritability, or apathy—when menu items become repetitive "junk

food" instead of a source of nourishment. Counterbalance them with restorative practices and variety.

7. **Use social "seasoning."** Involve trusted friends, mentors, or support networks to codesign or refresh your menu. Group challenges, constructive feedback, or shared celebration help keep you accountable and boost results.

8. **Prioritize intrinsic flavor.** Select activities and rituals that deliver meaning, growth, and connection—not just immediate pleasure. Center your menu on what aligns with your personal values and long-term aspirations, ensuring each "dish" is a step toward enduring satisfaction.

9. **Lean into compassion and curiosity.** Recognize setbacks and menu failures as part of the experimental process. Practice self-kindness, and remain curious about how new life stages or challenges may change your tastes over time.

Why this matters: With regular practice, your Dopamine Menu will become an enduring, adaptive guide—transforming every day into an opportunity to nourish motivation, enhance resilience, and cultivate a more profound sense of joy.

CHAPTER SUMMARY

A personalized Dopamine Menu turns a scattered routine into a dynamic practice of motivation, creativity, and emotional resilience. By auditing, curating, and adapting your menu, you forge a system that supports flourishing in all aspects of life. Here are the main takeaways to remember from this chapter:

- Individual menus and ongoing experimentation are essential to optimize dopamine and create true satisfaction and focus.
- A successful, motivating menu stems from regularly rotating items, striking a balance between challenging and restorative

practices, and timing menu items to align with your natural energy rhythms.

- Social connection amplifies the effects of your Dopamine Menu; shared rituals and collaborative challenges add motivation and accountability.
- Lifelong fulfillment and mastery of the dopamine system emerge from treating each routine as a living dish—removing, refreshing, and savoring options as you grow.
- Compassionate reflection can turn setbacks into opportunities to learn more about what works and doesn't work in your menu.

CHAPTER 9

Dopamine Detox

A Tool, Not a Cure

Dopamine detoxing has captured widespread attention for its promise as the ultimate break for overloaded minds and distracted lives. However, enduring change requires far more than a quick escape from stimulation. A helpful detox is not about abstaining or fasting as a way to shift patterns, but rather pausing to ground yourself for a more genuine, sustainable, and rewarding existence.

In this chapter, you'll uncover the neuroscience of the dopamine detox, its truths and misconceptions, and how to use this reset not as an end point, but as a launching pad for lasting growth. Why do constant cravings, distractions, and emotional flatness persist even after days of being unplugged from the Internet? A true detox works as an intentional interruption—a moment for you to examine, with more focused clarity, the ways digital, social, and habitual loops shape your brain's reward system. This intentional break, when used properly, can reveal hidden vulnerabilities and surprising sources of satisfaction, opening the door to more purposeful habits and deeper self-mastery.

But just hitting pause isn't enough. Ongoing progress demands a longer-term vision. In the following pages, you'll find science-backed guidance for planning, executing, and most critically, transitioning out of a dopamine detox. You'll learn to map triggers for reward-seeking behaviors and overstimulation loops, set achievable milestones, and create a fresh "palette" of experiences that become new main courses, specials, and restorative items in your Dopamine Menu. You'll discover firsthand that actual change grows not by avoidance, but through routines founded in meaning, anticipation, and lasting fulfillment. You'll also meet Andre, whose love for competition spiraled into a compulsion until he tried an effective and structured dopamine detox. Throughout this chapter, you'll explore practical approaches for building a more adaptable, enriching relationship with motivation—using biology, reflection, and mindful experimentation to create a sense of satisfaction and vitality that endures long after any detox.

"Resetting" Your Reward System: The Science and Strategy Behind the Reset

In an age saturated with engineered novelty and rapid-fire entertainment, it's no surprise that the idea of a dopamine detox appeals to anyone looking for more focus and clarity. In today's world, you are navigating landscapes that are nothing like those of your evolutionary ancestors; every beep, ping, and swipe delivers a brain jolt that was once reserved for deep satisfaction, real connection, or meaningful triumph. The tendency to chase stimulation isn't about weakness—it is a by-product of how your brain is wired to seek the unexpected and the exciting.

As you learned in Part 1, each habitual scroll, impulsive click, or craving for the next distraction forges neural pathways that keep you focused on what's next. Over time, your threshold for thrill rises. Activities that once delighted you can now seem muted, and your brain begins to crave ever greater stimulation. A thoughtful dopamine detox serves to interrupt these cycles of seeking more and more. By deliberately stepping away from digital, social, or habitual temptations, you create a space for your nervous system to recalibrate. Studies show that even short breaks from stimulation can help dopamine receptors become more sensitive—making small pleasures and ordinary moments feel vivid again.

This change reflects a broader shift in modern culture: More people are rediscovering the value of slower, deeper forms of meaning, such as meals enjoyed with friends, tactile creativity, soul-nourishing time in nature, and honest and present conversations. As artificial rewards like smartphone notifications quiet, your mind can better value the deeper payoffs of patience, learning, and the authentic presence—the same qualities that sustained your ancestors before constant distraction existed.

This renewal, however, is neither automatic nor effortless. During the early days of a dopamine detox, the nervous system may push back with annoyance, boredom, or even a vague sense of loss. Fortunately, these are clear indicators that change is underway. Recognizing these feelings as a growth opportunity rather than a setback can empower you to hold steady through the detox. The most lasting changes occur when you turn absence into curiosity: letting restraint serve as a path to discovery, and temporary discomfort as a step to a richer, more meaningful way of engaging with life. You'll uncover more about what to expect during a dopamine detox later in this chapter.

The length of a dopamine detox depends on your individual baseline and goals. Most people benefit from a four- to six-week detox to genuinely interrupt overstimulated reward pathways and begin recalibrating sensitivity. Some may need shorter two- to three-week resets to address acute overstimulation, while others with more significant dysregulation may need eight to twelve weeks. You'll know when you're ready to transition out of detox when mental clarity returns, focus sharpens, and simple daily moments feel genuinely engaging again. You'll have more authentic motivation to start and complete tasks without forcing effort; you'll notice an improvement in sleep quality, and you'll have emotional stability where irritability once dominated. The constant pull toward your phone will be significantly diminished; you'll find yourself reaching instinctively far less often.

Why "Just Stopping" Isn't Enough: Using the Detox Window Wisely

A pause from overstimulation—whether it's social media, junk food, gambling, or chasing constant success—can be a breath of fresh air. But research shows that hitting pause, by itself, is rarely enough to rewire old reward patterns. If the space you create in the detox isn't filled with intention and discovery, it is easy to slide back into past cycles, sometimes with even greater urgency. This is not a character flaw, but rather a reflection of how your brain is wired: When familiar excitements disappear but nothing nourishing takes their place, your innate quest for satisfaction will seek out what's easy, familiar, or even self-defeating to fill the gap.

The true power of a dopamine detox is in what you do with the breathing room it provides. Take time to revisit dormant joys, recall forgotten sparks of delight, and explore new forms of gratification that resonate with your current values. What pursuits genuinely leave you refueled? Where do you find a sense of authentic connection versus shallow affirmation? What shifts when you trade instant gratification for creative immersion, quiet solitude, or meaningful movement?

Some examples of impactful things you can do during detox include:

- Revisiting a long-neglected book or journal you've meant to explore
- Pursuing a skill you've been deferring, such as painting, woodworking, or musical instrument practice
- Taking extended walks in unfamiliar neighborhoods to rebuild observation and spatial awareness
- Having unhurried conversations with people you care about without devices present
- Making a meal from scratch and noticing the flavors and textures as you enjoy it
- Writing a letter to someone by hand
- Visiting museums or galleries
- Practicing meditation or breath work

Treat the detox as an experiment—a chance to recalibrate your inner sense of reward. Pay attention to both frustrations and small wins during this time. Eventually, you'll begin to uncover richer, more enduring motivators that quick fixes can never provide. Real progress lies not in simply avoiding distractions, but by intentionally choosing meaningful and vibrant rewards.

Performance, Well-Being, and Relationships: Real-World Applications

The dopamine detox is more than a way to disrupt unhealthy routines—it sparks a cascade of benefits that impact every part of your life. As compulsions fade and your brain's appetite for authentic reward returns, your mental resources shift from craving to focusing on richer, more intentional pursuits.

In Relationships

One of the biggest—and often overlooked—transformations of the dopamine detox happens in how people connect with each other. Without the clutter of screens and the need for instant validation, people begin to notice their loved ones in new ways: subtle body language, unspoken needs, and the unique rhythm of shared experiences. Couples often rediscover moments of playful intimacy; families build healthy rituals around meals, activities, or conversation; friends feel truly heard instead of merely seen. Social gatherings evolve, moving away from distraction toward authentic engagement—collaborating, laughing, learning, and building the resilience that makes real connection endure.

In Learning, Growth, and Self-Discovery

When your brain isn't chasing constant stimulation, it's easier to focus and do some deeper exploring. Reading, creative hobbies, or skill-building pursuits can give a sense of "flow," where you become so absorbed in an activity that time seems to disappear. Challenges feel less discouraging because your brain is better at adapting to setbacks, seeing them as opportunities for growth instead of something to avoid.

In Professional Settings

A dopamine detox promotes sharper concentration and a renewed ability to approach challenges at work with creativity, strategic vision, and depth. Without constant dopamine spikes, long-term problem-solving and innovative thinking come more easily. Those in leadership roles report greater adaptability, while complex projects feel less overwhelming, thanks to better focus and greater emotional flexibility. Professional relationships also tend to mature: meetings carry more substance, feedback becomes genuinely constructive, and group morale shifts from superficial busyness to purposeful, meaningful advancement.

In Physical Presence and Sensory Vitality

As cravings fade, your focus returns to your physical well-being. The diminished urge for distractions often comes with a renewed appreciation for nourishing food, restorative sleep, and regular physical activity. Walks can become meditative, flavors and aromas can regain complexity, and music or touch can be savored more deeply. Even the subtlest moments—a warm breeze, a melody, laughter—can become newly vivid. This renewed sensitivity shows that your brain is primed to experience the richness of the present moment.

Navigating Emotional Turbulence: Anxiety, Uncertainty, and Mood Shifts

Detoxing an imbalanced dopamine system often comes with unexpected emotions: waves of anxiety, restlessness, frustration, or even subtle loss. These reactions don't mean the detox isn't working. In truth, they prove that your brain and nervous system are recalibrating and learning how to search for fulfillment through deeper, healthier channels.

Neuroscientific research shows that adjusting reward sensitivity can lead to growing pains. While the release of dopamine during anticipation and pleasure may dip temporarily, this is part of your system's adaptation process—not a reason for alarm. From an evolutionary and social perspective, humans have always been wired to chase novelty and seek safety cues from the world around them, so the sudden absence of stimulating signals can naturally leave you feeling uncertain.

Instead of labeling these uncomfortable emotions as setbacks, treat them as valuable indicators. Temporary anxiety, longing, or unease may be prompting you to reconnect with values that you've been ignoring, explore forgotten sources of joy, or experiment with new coping strategies that don't rely on external stimulation. Often, discomfort is simply a prompt to rest, seek meaning, renew relationships, or move your body.

Mindfulness and gentle self-reflection can serve as powerful allies during these challenging moments. When anxiety or restlessness surfaces, pause and observe without judgment. Take note of physical sensations, thoughts, and feelings, and what deeper need might be guiding you. Reframe these situations as a chance to discover a healthier distraction: What new practice, environment, or companionship could offer comfort or pleasure right now? A walk in nature, an openhearted conversation with a loved one, or other restorative distraction can be the best medicine.

If you experience more persistent worry or difficult symptoms, reach out for support. Speaking with a trusted friend, a professional advisor, or someone else who is trying a dopamine detox can create a sense of shared resilience and new perspective. Each time you successfully navigate these emotional tides, you are not only building self-confidence but also training your brain to find contentment that lasts well beyond the detox period.

Ride each emotional wave as part of the process. Every challenge and insight contributes to a broader capacity for well-being, adaptability, and mastery of your dopamine system.

CASE STUDY
Andre—From Compulsive Thrills to Genuine Joy

Andre always saw himself as enterprising—a small business owner who thrived on competition and risk. But what began as the pleasure of poker nights gradually snowballed. Over time, Andre was chasing bigger highs: first through fantasy sports and betting pools, then through rapid-fire sports apps and mounting wagers. The thrill of a win quickly gave way to feelings of emptiness, agitation, and distracted thinking that bled into his work and strained his marriage.

When we began working together, Andre resisted the idea that his cravings reflected more than idle habits. He viewed them as quirks, not compulsions, and certainly not a matter of brain circuitry. Our first step was a "dopamine inventory"—tracking the patterns, cues, and emotional drivers behind his urges, along with the stressors that pulled him back to the table or screen.

We crafted a seven-day detox—no betting, no tracking sports scores, no "just checking the odds"—and paired it with daily reflection and simple, new rituals for moments when temptation struck. The first days of his detox brought irritability and a sense of loss, particularly during lulls or after high-stakes workdays. Yet, as Andre consciously faced these urges by naming his cravings, charting their intensity, and choosing activities such as brief walks, supportive phone calls, or hands-on projects instead of gambling, his definition of satisfaction began to shift.

Midway through the detox, Andre reported subtle but remarkable changes: He slowed down to enjoy breakfast, felt a true sense of accomplishment after a challenging project, and experienced renewed conversation and laughter with his wife. He saw clearly how the constant chase had blunted his capacity for authentic pleasure. By the end of the week, Andre hadn't erased his struggles, but he had discovered agency. With a new awareness of his triggers, we built a Dopamine Menu that included lunch dates, outdoor runs, creative business planning, and old-fashioned board games. Each activity provided Andre genuine connection and lasting contentment.

Andre's journey illustrates a key principle supported by neuroscience: The power of a dopamine detox isn't in deprivation, but in using the reset to find better sources of engagement, connection, and self-worth. True satisfaction emerges not from higher stakes, but from reclaiming the simple, meaningful joys of daily life.

The Exit Strategy: Transitioning Out of Your Detox with Intention

Coming out of detox requires as much intent as the detox itself. This isn't a space to simply resume old patterns or revert to your previous baseline. Instead, treat it as a guided reintroduction where you systematically restore specific elements while monitoring your nervous system's response.

Begin by reintroducing activities and relationships first. The activities you engaged in during your detox—the items from your Dopamine Menu that proved genuinely rewarding—become your foundation. Return to real-world social connection before digital connection. Schedule in-person time, phone calls with close friends, and collaborative projects that require presence. These rebuild social reward circuitry in ways that feel nourishing rather than compulsive.

Next, strategically restore technology and specific platforms. Rather than returning to everything at once, select one platform or tool and reintegrate it consciously over three to five days. Observe how your energy, sleep, and motivation respond. Are you reaching for the phone instinctively? Do old compulsions resurface? Use these signals as important indicators. If regression occurs, extend that item's absence or reconsider whether it belongs in your menu at all.

Reintroduce high dopamine activities last, and only those aligned with your menu. If streaming felt like a compulsive escape before, perhaps it doesn't return to your daily rotation. If certain social media platforms triggered comparison spirals, maybe they should remain offline. Be selective. Your detox revealed what genuinely served you versus what captured your attention through design manipulation. Honor that clarity.

Throughout this transition, maintain the practices that supported you during your detox, whether it's meditation, journaling, or extended time outdoors. These aren't temporary scaffolding; they're foundational to long-term dopamine balance.

Continuing the Journey

Finishing a dopamine detox is not the end of struggling with distraction and craving. Instead, it is a springboard to ongoing growth and deeper self-awareness. Real resilience develops not simply by resisting temptation, but by embracing new habits and perspectives so that the pull of old pleasures gradually loses its grip. In its place, authentic rewards become the default in your dopamine pathways. Picture your mind as a greenhouse: Regularly "pruning" compulsive behaviors, planting "seeds" of novel interest, and consistently tending to the "blooms" through reflection all help cultivate a balanced, thriving ecosystem. Scientific studies confirm that each deliberate choice—whether toward a meaningful project, a new relationship, or genuine rest—rewires your brain, making fleeting thrills less appealing and life's deeper satisfactions more prominent.

Treat "menu revision" as an evolving ritual. Schedule weekly or monthly reviews to notice which activities spark energy, which ones slip into monotony, and which might need fresh experimentation. Celebrate small discoveries—a recipe enjoyed, a tech-free walk, or the revival of an old friendship. Noting progress, no matter how small, strengthens the pathways for anticipation and gratification and sets the stage for better choices tomorrow.

As you tend to your own inner greenhouse, here are some helpful tips for making the most of a dopamine detox:

- **Structure your space to support healthy patterns.** Put phones out of reach, display visible reminders of your ambitions (a handwritten list of goals pinned above your desk, a vision board capturing the person you're becoming, or notes on your mirror about what matters most); you can also invite others

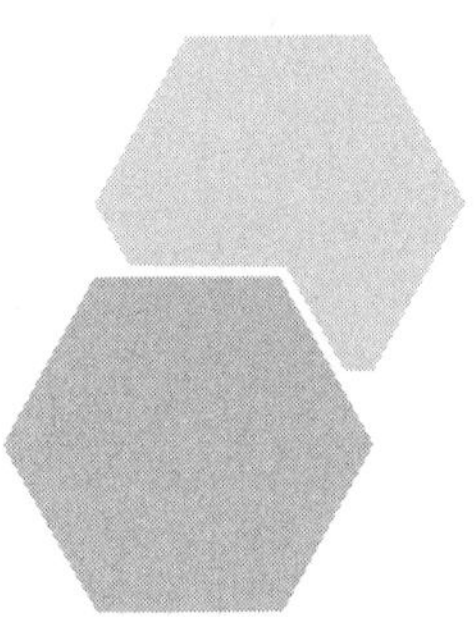

Real resilience develops not simply by resisting temptation, but by embracing new habits and perspectives so that the pull of old pleasures gradually loses its grip.

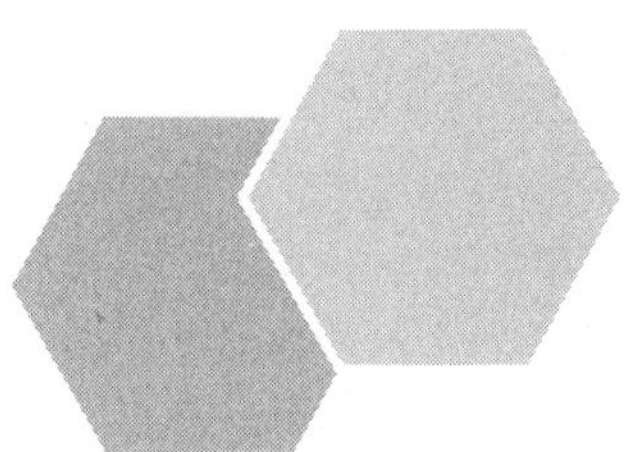

to join the detox. Just as your ancestors thrived by weaving adaptive routines into their changing surroundings, these tiny environmental tweaks can serve as gentle guides.

- **Reward effort alongside outcome.** Acknowledge not just significant breakthroughs, but every time you resist a craving, venture into something novel, or simply the fact you are pausing to reset. These micro-rewards, grounded in self-encouragement, can help keep you motivated to create lasting change.
- **Anchor your mindset in self-compassion.** Slumps, relapses, or moments of temptation are natural feedback from your brain—not markers of failure. Each renewed commitment and fresh experiment is what shapes resilience and personal insight.
- **Revise, revise, revise.** A detox is as adaptive as your Dopamine Menu, so be sure to routinely check in with how it is going. Adjust activities and adapt to the changes as you detox.

By living out the rhythms of the dopamine detox—revising your practices, choosing rewarding activities, and honoring each stage of progress—you build a system where thriving is steady, personalized, and restorative.

Practical Strategies: Your Dopamine Detox Playbook

Pause and practice: Rewiring your brain's reward system takes more than avoiding temptation; it's a deliberate, structured process. Use this space to practice what you have discovered in this chapter. The following are the steps for detoxing from dopamine imbalance:

1. **Prep for success.** Start by identifying the habits or environments that most often hijack your attention and push you to constantly seek new stimulation. List which digital platforms, substances, foods, or routines feel hardest to resist. Rather than focusing

on shame or self-blame, adopt the mindset of a scientist: This is simply data collection. Define a clear intention: What do you hope to notice, feel, or learn by creating this detoxing space?

2. **Establish boundaries.** Choose a realistic length for your detox, whether it's one weekend, a workweek, or an extended period. Remove obvious triggers from your environment (such as phone apps and notifications, streaming subscriptions, snacks that encourage browsing breaks, etc.), and let others know you're taking a pause. This both reduces temptation and adds a gentle social accountability.

3. **Replace rather than remove.** Sketch out alternative routines for when you feel temptation, such as reading or writing during screen breaks or taking outdoor walks in lieu of playing online games.

4. **Embrace discomfort as feedback.** Expect waves of restlessness, nostalgia, or longing. Remember, these aren't signs of failure; they're the natural response of an adjusting nervous system. Use this discomfort as a cue to journal, observe patterns, or experiment with new mini-rewards. Notice which moments are most challenging and which activities bring a surprising spark of pleasure or calm.

5. **Curate new "menu items" during your detox.** Use this time to explore experiences you once ignored for fast rewards—try manual projects, in-depth conversations, books, movement, or cooking. Let curiosity and sensory satisfaction guide you. Ask yourself, "What does my brain truly crave now?"

6. **Reflect, review, and reinstate with intention.** When your chosen detox period ends, resist the urge to rebound immediately to old patterns. Instead, do a gentle review of the experience: which cravings faded, what unexpected satisfactions appeared, and what routines felt essential for well-being? Use your findings to adapt your Dopamine Menu so it always offers a dynamic selection of habits, relationships, and rewards that boost energy, motivation, and satisfaction. Commit to regular audits through structured reflection: Consider journaling your dopamine choices weekly,

and review which menu items genuinely served you and which made you slip into compulsion. Also take note of seasonal or life stage shifts that call for menu recalibration. Doing this will allow new habits to deepen; remember that slipups can become opportunities for growth. (These audits don't require returning to full detox; they're gentle check-ins that keep your menu alive and responsive to your evolving self.)

Why this matters: By following this process of detox, reflection, and renewal, you use the science of dopamine and your brain's natural adaptability. Over time, you turn a break from bad habits into lasting happiness and achievement.

CHAPTER SUMMARY

The dopamine detox is a launching point rather than an end point—an intentional reset that sparks lasting resilience. Keep this chapter's main takeaways in mind as you continue onto the next:

- The dopamine detox reveals which cravings are compulsive loops versus which activities genuinely nourish you.
- Real resilience emerges through new habits and perspectives that gradually shift old pleasure pathways, making authentic rewards your brain's default setting.
- Structure your environment deliberately by removing obvious triggers, displaying visible reminders of your values and ambitions, and enlisting social accountability to support the detox.
- Acknowledge effort alongside outcomes by recognizing every step of resistance, experimentation, and self-reflection. These mini rewards strengthen anticipation pathways and build momentum for lasting behavioral change.
- Practice self-compassion as you detox, treating setbacks as natural feedback rather than markers of failure.

CHAPTER 10

Dopamine and the Body

Movement, Nutrition, and the Biochemistry of Motivation

Motivation is not a miraculous force; it is built into your biology, within flesh and bone, shaped by ancient circuitry that evolved to sense, seek, and adapt. At the center of this is dopamine: the silent conductor turning anticipation into momentum and intention into movement. Sunlight on your skin, flexing your muscles in a stretch, taking in the scent of food, the true rest found in deep sleep—these are not simply "healthy habits," but life's original motivators. Each small physical act is fuel for the brain's reward system, triggering a release of energy, sharpening attention, and nudging desire toward growth instead of routine.

This chapter explores the anatomy of motivation. You'll learn about how dopamine rises and falls with your waking and sleeping patterns, and how it works alongside hormones like melatonin to set the pace for energy and rest. You'll also meet Luca, who was trapped in a cycle of late nights and erratic eating until he started honoring his body's needs. When sleep, movement, or eating habits get thrown off, these changes can ripple through your brain's feedback systems, quietly draining motivation and making everything feel like an uphill climb. As you'll read about more in this chapter, simple moments, such as stepping into sunlight or eating at regular times, will help your body tune into these rhythms, keeping motivation steady and grounded in ways you will deeply feel.

Today's world often promises quick fixes for fatigue and a lack of drive: energy drinks, late-night screens, and perfectionistic plans that never really deliver a satisfying boost. Understanding motivation on a biological level opens a powerful new path to maintaining drive every day. By mastering the subtle inputs of light, movement, food, and rest, you can harness motivation that goes beyond fleeting willpower to sustained momentum. Here, the real architecture of drive is revealed: a living dialogue between habits and neurochemistry, where every day holds new opportunity for success in your goals.

Movement: The First Messenger of Motivation

Movement is the oldest language of your brain and body; long before words, stories, or even conscious thoughts occurred in human history, there was action. Evolutionarily, your ancestors needed to move in order to find food, shelter, and community. Therefore, the body adapted so that every voluntary step, stretch, or gesture sent signals through the nervous system that boosted mood, focus, and sense of agency. Movement isn't just exercise; it is how your body communicates that it is ready for life and helps your brain see challenges as opportunities rather than threats.

Every time you move your body, your brain releases a mix of chemical messengers: Dopamine plays a leading role here, but it works with endorphins, serotonin, and even growth factors, which help your neural circuits adapt and thrive. Simple motion—standing up, walking into sunlight, reaching for a glass of water—signals the central nervous system to be alert and focused. Over time, these signals accumulate, shifting your baseline from tiredness to engagement. This isn't just about movement in the conventional sense, either. Vigorous activity—such as running, dancing, or resistance training—can dramatically boost dopamine and serotonin, creating a lingering sense of purpose and reward long after you've stopped exercising. In today's world, you can honor the evolutionary roots of activity by using intentional movement as a tool to shake off inertia and unlock drive.

Movement also restores the body's sensitivity to pleasure and novelty, making every experience—whether work, rest, or social interaction—more vivid and rewarding. It is no surprise, then, that some of the world's most creative breakthroughs occur not at a desk, but during a walk, a workout, or while tending to basic physical needs. By treating movement as an active partner in motivation, you can use every step, stretch, or dance as a spark for meaningful, enduring change.

Nutrition: Fuel for the Neurochemical Engine

In addition to how (and how often) you move, what you eat affects how your brain releases and uses dopamine. Beyond calories or flavor, the act of feeding your body real, diverse foods gives it the raw materials needed to produce molecules such as dopamine and other helpful chemicals. When meals are balanced and thoughtfully timed, they sharpen attention, steady energy, and promote resilience to both stress and temptation.

Dopamine relies on certain nutrients, including amino acids, vitamins, and minerals, which are found in everyday foods. Tyrosine, for example, is a fundamental amino acid present in eggs, beans, fish, and seeds; it's converted within the brain into dopamine, supporting drive and emotional steadiness over long periods of time. The brain is also impacted by blood sugar levels; frequent spikes and crashes in blood sugar due to processed foods can hijack dopamine pathways, leading to changes in mood and motivation that are difficult to control consciously.

Evolution did not design humans for artificial flavors or overeating. Instead, real food—fresh vegetables, fruits, grains, and high-quality proteins—balance out the brain's reward system and produce the chemicals for drive and satisfaction. Shared meals, mindful eating (eating with full attention to the sensory and emotional experience of food, noticing flavors, textures, and aromas without judgment or distraction), and consistent nourishment is more than cultural tradition; it's crucial to the brain's sense of safety, reward, and renewal. In contrast, erratic food patterns like chronic dieting, skipping meals, or late-night bingeing can erode the brain's natural balance, leading to cycles of exhaustion and craving.

When it comes to nourishment, modern life offers endless convenience—but also relentless temptation. Highly engineered snacks and drinks provide short bursts of pleasure but rarely satisfy the deeper hunger for stable energy and clearheaded motivation. By understanding nutrition as a constant conversation between the chemicals of the brain, you can start to build a foundation for empowered nourishment. Each bite becomes an act of partnership with your body's ancient wisdom—a daily choice to support the mental and physical vitality needed to pursue your goals.

Sleep: The Unsung Guardian of Motivation

Sleep is often the first thing to be sacrificed when life gets busy, yet it is key to every spark of motivation you have. While modern society frequently glorifies hustling and late nights at the office, science continues to prove that restorative sleep is a biological necessity; it radically shapes every layer of drive, resilience, and well-being. Each night, as you cycle through **light**, **deep**, and **rapid eye movement** (**REM**) **sleep**, your brain recalibrates its dopamine systems, allowing you to reset your natural appetite for reward, focus, and anticipation for the day ahead.

Sound, uninterrupted sleep is essential for keeping dopamine in balance. Without enough deep sleep (a recommended seventy-two to ninety-six minutes per night for a typical eight-hour sleep), dopamine production, signaling, and receptor sensitivity can all be disrupted. This shift doesn't just make you feel groggy—it drains motivation. When sleep is consistently shortchanged, dopamine's natural ebb and flow falters. Suddenly, you can find yourself struggling to take healthy risks or try again after setbacks. Even small daily tasks can feel overwhelming when dopamine is out of balance.

A lesser-known, yet important, fact is that sleep loss can also make it difficult to regulate emotions. Anger, frustration, and impatience become more pronounced, not solely because you are tired, but because your brain's regulation of emotion and impulse control are controlled by dopamine. Chronic sleep loss can make you more prone to reacting quickly and less able to recover from everyday stressors.

A CLOSER LOOK
Light, Deep, and REM Sleep

Your brain cycles through four distinct sleep stages, each critical to dopamine regulation and motivation:

Stage 1 (Light Sleep): The transition between wakefulness and sleep. Your body begins to slow. This stage lasts one to seven minutes and happens multiple times per night.

Stage 2 (Light Sleep): Your heart rate slows, body temperature drops, and brain waves slow further. This is where memory consolidation begins. You spend about 45%–55% of your sleep in this stage.

Stage 3 (Deep Sleep): Also called slow-wave sleep, this is where physical restoration happens. Your body repairs tissue, builds muscle, and releases growth hormone. This is the sleep your dopamine system needs most for resilience. You spend about 12.5%–20% of total sleep here.

Stage 4 (REM Sleep): REM sleep is where your eyes move rapidly under your eyelids and your brain operates at near-waking levels of activity. During REM sleep, your brain integrates the day's learning and experiences into long-term memory, particularly emotional and social memories. REM sleep is also where your dopamine system processes reward and motivation. You spend about 20%–25% of sleep in REM.

You cycle through all four stages multiple times per night, each complete cycle taking about ninety minutes. Skipping deep sleep or REM disrupts dopamine regulation and leaves motivation fragile.

Restorative sleep protects
the dopamine system,
allowing you to make clear
decisions, solve problems,
and more steadily respond
to challenges.

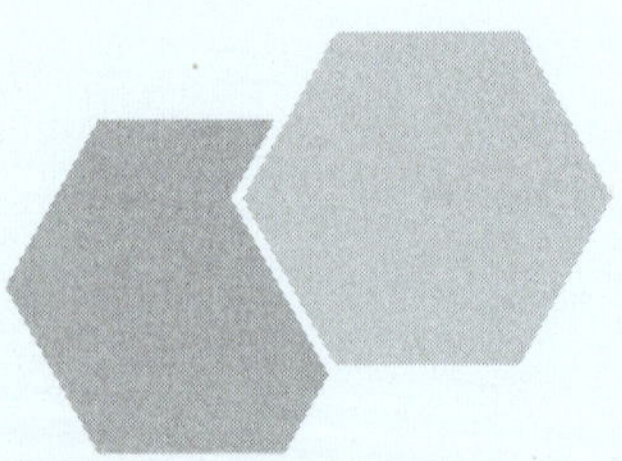

Restorative sleep protects the dopamine system, allowing you to make clear decisions, solve problems, and more steadily respond to challenges. Returning to regular bedtimes, dimming evening lights, and winding down with soothing rituals can help reset natural **circadian rhythms**, allowing dopamine to do its best work.

A CLOSER LOOK
Circadian Rhythm

Your circadian rhythm is regulated by light: Exposure to morning light signals your brain to wake up; evening darkness signals the production of melatonin, the hormone that promotes sleep. When your circadian rhythm is synchronized, you sleep and wake at consistent times and your dopamine system operates at peak efficiency. When the circadian rhythm is disrupted through irregular sleep patterns, late-night blue light exposure, or work schedules that require waking and sleeping at unconventional times (such as night shifts or rotating schedules), dopamine signaling becomes erratic. Your brain struggles to produce motivation reliably, and mood dysregulation follows.

When sleep is honored, motivation and enthusiasm return, movement feels more natural, and setbacks are easier to manage. At its core, sleep is more than rest; it's a renewing ally in focus, creativity, and drive.

How Dreams Rehearse Motivation

People often think of sleep as a time when the brain simply powers down, but neuroscience reveals that it's actually very active—especially when dreaming. While you sleep, your brain's reward systems, including dopamine pathways, come alive, particularly during REM sleep and deep, restorative sleep. During REM sleep, which happens about ninety minutes after you fall asleep, the areas of the brain linked to desire, exploration, and drive light up, as if your mind is practicing action and decision-making while you dream. Memory researchers have discovered that when dreams include elements of real-life learning and working toward goals, you are more likely to remember and act on those lessons when you wake up. In this way, dreaming helps reshape how motivation "feels" in your brain—connecting it to curiosity, anticipation, and emotions. It's through these connections that you feel the drive to move toward what matters.

Dreaming also strengthens the circuits related to reward, novelty, and even risk-taking. When dreaming is rich and varied, especially following days of challenge or growth, people often wake up with a fresh perspective or renewed enthusiasm. Some neuroscientists compare this to an internal "rehearsal," where motivation is fine-tuned and goals are adjusted with an emotional "charge" for the days ahead.

What does this mean for daily life? Simply put, the quality and content of your dreams aren't just trivial stories; they are blueprints for motivation. Prioritizing sound, uninterrupted sleep not only boosts energy and mood, but it also allows the brain's dopamine systems a chance to rehearse, adapt, and ultimately empower your sense of drive. As science continues to uncover more about this nightly rehearsal, one thing is clear: The foundation for tomorrow's motivation is found not just in what you do today, but in the restorative, imaginative, and dopamine-rich worlds of your dreams.

Sunlight and Circadian Rhythm: Nature's Blueprint for Motivation

Long before alarm clocks and digital screens, the rising and setting of the sun governed human energy and drive. The body and brain adapted to daily cycles of light and darkness—cues that you still depend on today to regulate feelings of alertness, focus, and motivation.

Exposure to natural light in the morning signals the brain to release a potent mix of chemicals, including dopamine, serotonin, and cortisol, that lift your mood and prepare you for movement and engagement. This boost acts as a daily reset, signaling to your body that it's time to go from rest to action and ensuring that energy, creativity, and motivation are aligned with your daily to-dos. Sunlight also helps keep your circadian rhythms sharp, allowing your body to anticipate patterns of activity and rest. The ability to anticipate your daily patterns is important because this predictability reduces chronic stress, stabilizes cortisol, and allows you to feel ready for what's to come.

Conversely, missing out on natural light or exposing yourself to artificial light late in the evening can throw off these internal rhythms. The result can be a feeling of fogginess in the morning, evening restlessness, or even a creeping sense of stagnation. Over time, disrupted circadian signals can erode motivation, silencing the brain cues that would otherwise trigger positive anticipation. Screens and indoor routines, while convenient, often dim the nervous system's capacity to reset fully, leaving the body chasing energy and focus that never quite arrive.

The good news: Small, intentional shifts make a dramatic difference. Stepping outside during the first hour of daylight, pausing work to take a walk or stretch under an open sky, and dimming lights in the evening all help reinforce the natural ebb and flow that supports motivation. Working with light instead of against it improves alertness and boosts confidence and balance.

Motivation's Paradox: Why Drive Can Strengthen or Collapse

Stress often gets a bad reputation, but it isn't always harmful. In reality, stress has played an important role in human survival for thousands of years. The body's automatic stress response acts as an alarm system, focusing attention and boosting energy when an actual threat pops up. Short bursts of stress can temporarily heighten motivation, sharpen resolve, and prompt quick action when it matters most. In small doses, stress can be motivating, helping you learn, adapt, and push through moments of uncertainty.

Yet, the line between stress and motivation is thin. When stress becomes chronic or overwhelming, it shifts from a valuable tool to a powerful saboteur. Prolonged stress interrupts the delicate balance of dopamine in the brain's reward and motivation pathways. What was initially energizing eventually becomes draining, sapping enthusiasm and making it harder to care about goals or new challenges.

Biologically, the effects of long-term stress are unmistakable. When stress hormones such as cortisol stay high for too long, dopamine's healthy cycles are interrupted. With time, this disrupts the pathways that regulate pleasure, anticipation, and reward. Tasks that once seemed inviting can start to feel overwhelming. When this happens, the mind gravitates toward habitual or numbing behaviors (which seem easier) instead of intentional, purposeful action (which seems more effortful).

Not all stress impacts motivation equally. The way stress is experienced—how intense it is, how long it lasts, and whether you feel a sense of control during it—determines whether it nudges you forward or sets you back. Ultimately, stress is a double-edged sword: It can motivate you or act as a barrier to progress. The key is learning how to manage it so that it works *for* you, not against you. Understanding how stress affects dopamine helps you tell the difference between energizing challenges and harmful overload. Building steady routines, setting boundaries, and protecting rest all support a more balanced, motivated mind.

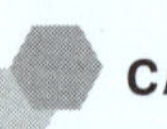

CASE STUDY
Luca—Restoring Drive One Ritual at a Time

Luca seemed like a successful culinary professional working as a chef, but beneath the surface, he was struggling to feel motivated and enthusiastic. Years of late-night shifts, constant stress, and waiting to eat meals after work had created a pattern he struggled to break. Most nights ended with him raiding the kitchen and eating in silence after closing before falling into restless sleep. Mornings felt like a battle; even the smallest tasks seemed daunting, and the spark that once fueled his creativity in the kitchen was fading.

In our early sessions, Luca described feeling trapped by his own routines. He blamed himself for what he saw as a lack of discipline and willpower, yet nothing he tried made a sustainable difference. He wanted to feel a renewed sense of purpose in his work and relationships but felt stuck on a treadmill of exhaustion, cravings, and emotional flatness.

Our work began with a step back—not toward more rigid rules, but toward understanding how his biology and environment were shaping his experience. Together, we unpacked the role of movement, light, food, and sleep in regulating his motivation and emotional steadiness. Instead of overhauling everything, we started with a few small shifts: getting outside for sunlight within the first hour of waking, redefining late-night eating as a mindful and nourishing ritual rather than a guilty pleasure, and reestablishing a simple bedtime routine.

Gradually, Luca noticed changes that surprised him. Sunlit mornings and daily movement began to lift his energy and give him a sense of focus. More balanced and intentional meals, both at work and home, reduced cravings and helped him feel full throughout the day. And by honoring sleep with regular wind-down rituals, he found that rest became restorative again, and with it, his mood, patience, and motivation lifted. Interactions with colleagues became easier, he felt inspired in his cooking, and setbacks no longer spiraled into days of frustration.

Luca's journey wasn't about a single breakthrough or superhuman willpower. It was about relearning the natural conversations between body and mind and trusting that, with the right signals, motivation would return. His story is a reminder that the path to sustained drive is not through self-criticism, but by nurturing the everyday foundations of a vibrant, meaningful life.

Practical Strategies: Boosting Your Dopamine Menu

Pause and practice: Turning insight into action means translating science—and Luca's journey—into tangible, everyday practices. The following strategies are designed to help you make your own shifts to boost motivation and energy. Start where you are, adjust for your needs, and revisit these foundations whenever you feel drive starting to fade.

1. **Jump-start with sunlight.** Aim to step outside within an hour of waking, even if it's just for five minutes. Natural light triggers

the brain's dopamine and circadian systems, setting the tone for alertness and drive throughout the day.

2. **Move early and often.** Integrate movement into your morning and repeat short bursts of movement during the day—a few stretches, a brisk walk, or light exercise. Each small act of motion refreshes dopamine circuits, defeats inertia, and primes your mind for purpose.

3. **Nourish with intention.** Plan out balanced meals ahead of time, focusing on dopamine-supporting foods—lean proteins, leafy greens, fermented foods, nuts, seeds, and colorful vegetables. Avoid skipping meals or grazing late at night to maintain steady energy and mood and curb cravings.

4. **Treat sleep as nonnegotiable.** Create a ritual for winding down at the end of the day and keep a consistent bedtime. Dim the lights at least thirty minutes before sleep, avoid screens, and make rest a priority. Notice how your motivation shifts after just one night of full, restorative sleep.

5. **Honor your biological rhythms.** Pay attention to patterns that leave you feeling energized or drained. Adjust your schedule, light exposure, and activities to fit your natural peaks and valleys rather than forcing rigid routines that don't serve your body.

6. **Reflect and refine.** At the end of each week, spend a few minutes journaling about your experiences with clarity, energy, or enthusiasm that week, as well as times when inertia or frustration crept in. Use what you learn to tweak your approach and anchor new habits.

Why this matters: There's no perfect formula for motivation, but grounding your Dopamine Menu in light, movement, nutrition, sleep, and attention to your natural rhythms helps you move toward your goals. Over time, these small shifts become the backdrop for renewed enthusiasm, creativity, and grounded resilience.

CHAPTER SUMMARY

Understanding and shaping motivation involves caring for your body's basic needs. Keep these important lessons from this chapter in mind as you continue reading:

- Drive is sparked and sustained not by abstract willpower, but by daily practices anchored in movement, nutrition, restorative sleep, and exposure to natural light.
- Dopamine is the conductor of motivation, coordinating desire, anticipation, and willingness to pursue meaningful goals. Disrupted sleep, stress, or unhealthy habits can tip this system out of balance.
- Dreaming and quality sleep don't just restore the body; they allow the brain to "rehearse" drive at a subconscious level, supporting resilience and creativity.
- Real-life stories (like Luca's) demonstrate that reclaiming drive isn't about adhering to perfect routines; it's about carefully listening to and supporting the body's deep, evolutionary needs.
- Building a personalized Dopamine Menu means honoring biology: Consistently tune your habits, environment, and mindset to gain attention, effort, and possibility.

CHAPTER 11

Designing Motivation

How to Make Progress Automatic

The best experiences in life aren't found in chasing endless excitement or simply striving for balance. Rather, they are found in a genuine sense of purpose. What gives you the drive to not just look for, but achieve this purpose? Motivation. As you've learned in this book, motivation is one of the key products of a healthy dopamine system. While there are prescribed routines and exercises you can use to make your dopamine pathways work for you (as you explored in previous chapters), you can take things further by activating your brain's innate rhythms so motivation and growth come more automatically. Natural surges of enthusiasm, moments of contemplation, leaps of creativity, and periods of deep rest can all be worked into your personal Dopamine Menu for a life anchored in meaning.

In this chapter, you'll discover why satisfying achievement and lasting contentment occur when you adhere to your own personal values and conscious intentions—not by resisting your impulses. Neuroscience reveals a profound truth: By embracing your brain's cycles of anticipation and fulfillment, you can access reserves of ingenuity, fortitude, meaningful relationships, and inner joy. As you read through the following pages, you'll discern the difference between fleeting urges and the call of true devotion—your key for redirecting scattered energy into purposeful engagement. Through real-world examples and actionable guidance, you'll transform sparks of motivation into lasting habits, cultivate environments and rituals that boost concentration and emotional resilience, and design your daily life around your evolving goals.

This chapter is filled with practical insights for harnessing the natural cycles you learned about in Chapter 10. You'll learn how to nurture happiness in the present and foster a well-being that endures through the ups and downs of life. And you'll tap into real-world experiences through Mei, who felt paralyzed by procrastination and self-doubt until she took charge of her physical surroundings. Whether you're seeking greater momentum at work, stronger bonds at home, or progress through shifting terrain, the steps to automating your dopamine system are here.

The Dopamine Conductor: Moving Beyond Routine Toward Purpose

Before digging into the science of motivation and how you can make motivation feel more natural and automatic in your everyday life, it's helpful to look closer at purpose itself. Contemporary studies suggest there are four "batons" of purpose that ignite dopamine and make people feel motivated:

1. Intentional rituals
2. Diverse learning
3. Creative risks
4. Meaningful social ties

Those who thrive are known to intentionally weave these four elements of purpose throughout their days. They alternate periods of deep effort, reflection, relationship, and novel experience to turn effortful, dull routine into meaningful, vibrant living.

By understanding these basics of purpose, you can see how the invitation in this chapter isn't to discard the more automatic parts of living. Rather, it's to use these comforts as a springboard. By elevating everyday choices from simply involuntary to purposeful, you bring the full power of your reward system into alignment with a life that is deeply motivating.

Harnessing Anticipation and Recovery: The Neuroscience of Momentum

At the heart of human motivation is a balancing act between looking forward and letting go. The anticipation circuit in your brain transforms even ordinary goals into engaging pursuits. **Dopamine priming** (or engaging dopamine in anticipation of a big moment) isn't mere excitement. It helps explain why awaiting a holiday, the ritual of that first morning sip of coffee, or the nervous thrill before making your voice heard on a conference call can all propel progress. Neuroscientists confirm that dopamine peaks not when you receive a reward, but in those suspenseful moments of expectation and possibility that come before it. These moments are where motivation, curiosity, inventiveness, and bold choices come alive.

A CLOSER LOOK
Dopamine Priming

Dopamine priming is the strategic activation of your dopamine system before high-stakes moments. It allows you to sharpen your focus, motivation, and cognitive precision when you need it most. Unlike a dopamine detox (which resets your baseline) or dopamine fasting (which removes stimulation), dopamine priming is activational.

Yet, predictability weakens this effect. Research finds that lottery winners often report less day-to-day happiness than individuals who are making progress toward challenging, unpredictable goals. In animal and human studies alike, working toward inconsistent rewards or "occasional victories" builds more persistence and learning than routine outcomes. The lesson? Motivation pathways thrive when life is mixed with both defined aspirations and an element of surprise. Leaving space for novelty as well as challenge amid your goals is what helps you stay the course.

Equally essential to the element of surprise is how you **downshift** (or decrease mental and physical exertion) after periods of intense effort. True resilience doesn't mean nonstop striving; it involves knowing when your brain (and body) needs a rest. Scientists studying elite performers found that their most significant gains were not made by grinding endlessly, but by deliberately choosing activities such as unhurried walks, playful curiosity, and restful rituals that signal their brains to replenish and rewire after a challenge. Neuroscience continues to reveal that these transitions from action to rest activate new connections and solidify what's been learned, making each pause a hidden engine of future achievement.

A CLOSER LOOK
Downshift

Downshifting allows your dopamine system to recalibrate after a big push. Without downshifting, your baseline dopamine levels reset lower than usual and true recovery is more difficult.

Everyone can experiment with their natural cycles of momentum by trying these practices:

- **Unplug and reconnect.** Turn off alerts on all devices for periods of focused work, then restore yourself with meaningful movement (such as a reflective walk) or heartfelt exchanges (such as a vulnerable conversation with someone you love).
- **Design for anticipation.** Create a sense of excitement for significant milestones (for example, you can announce the milestone publicly to colleagues or your team). Set a specific reward tied to completion that engages your senses (such as a meal at a new restaurant, a weekend trip, or a personal

investment). Visualize the completion before you begin the effort, allowing your dopamine system to fire in anticipation.

- **Incorporate restorative activities that help ideas and energy settle in.** For example, journaling after completion of a project allows your mind to consolidate what was learned without continuing to problem solve. Meditative movement such as tai chi or gentle yoga transitions your nervous system from **sympathetic** activation (the fight-or-flight stress response) back to **parasympathetic** calm (the rest-and-digest state), embedding the learning into your body's memory. Creative play unrelated to your work (painting, music, gardening) resets your cognitive patterns and allows novel neural pathways to form around what you've accomplished.
- **Ride the rhythms.** Tune into your own, often unpredictable, tides of drive and recovery, and adapt rather than struggle against them.

Those stretches that seem "unproductive"—daydreaming, chatting, or simply being in the present moment—activate the default network in your brain tied to memory, creativity, and emotional processing. By balancing effort and renewal, you transform fleeting bursts of happiness into a sustainable, adaptive approach to success and well-being by making things more naturally motivating.

The Orchestra of a Fulfilling Life

In Chapter 1, you learned about how the dopamine system is like an orchestra. On a larger scale, your entire life can be viewed through this lens. Modern existence often steers people into familiar loops—predictable habits, digital diversions, small comforts—but the true "symphony" of fulfillment comes from actively shaping each part of the ensemble. In this process (or "performance"), dopamine acts as the concertmaster, driving anticipation, motivation, and vibrancy—it may not drive the entire score, but it keeps the performance moving.

Motivation pathways thrive when life is mixed with both defined aspirations and an element of surprise.

Leaving space for novelty as well as challenge amid your goals is what helps you stay the course.

Science shows that lasting satisfaction doesn't come from suppressing cravings or dulling emotions. Instead, the most resilient and inspired individuals are those who work *with* their dopamine system, intentionally guiding it and giving it the space to act as the concertmaster in their lives. Studies have shown that high-achieving athletes and inventive leaders have **reward agility**; this means that their brain circuits of expectation and pursuit aren't limited to immediate pleasure but also serve larger missions, deeper relationships, and journeys of self-discovery.

A CLOSER LOOK
Reward Agility

Reward agility is your ability to find motivation and satisfaction through varied reward pathways rather than relying on a single source of dopamine activation. An executive with high reward agility finds fulfillment not just in closing deals but in mentoring, creative problem-solving, or simply completing focused work. This flexibility prevents your body from getting so used to one reward type that it stops being as effective. It also helps you maintain sustainable motivation across different areas of your life.

It is purpose, not fleeting delight, that fine-tunes the brain for growth and lasting joy.

Designing Your Environment for Dopamine Mastery: How Your Surroundings Shape Your Neurochemistry

Motivation doesn't come just from inner resolve. As modern behavioral science reveals, it is strongly shaped by your surroundings. Every aspect of your physical and digital setting—lighting, arrangement of space, proximity to your work essentials versus proximity to distractions like food and social media—works quietly to either amplify or dampen your brain's dopamine pathways. Motivation, and the progress it depends on, is less a matter of sheer will and more a product of environments that support focus and follow-through.

Recent studies demonstrate that you can make big differences simply by shifting visible triggers: Relocating electronic devices, optimizing workspaces for immersive flow, and establishing distinct zones for relaxation can fine-tune your dopamine activity. Mindfully adapting your surroundings, whether at home or in an office, lays the groundwork for consistent momentum and a more rewarding daily experience.

The following are tips for using environmental cues and **friction** (or engineered resistance) to subtly and creatively boost motivation:

- Make symbols of good activities visible, such as books, journals, or workout equipment.
- Create gentle barriers for those habits you'd like to reduce, such as muting apps or placing junk food in less accessible locations.
- Break larger ambitions into small, tangible steps.
- Use vibrant reminders (such as a colorful vision board on your desk, sticky notes with motivational phrases in bright colors, or a progress tracker with visual markers you update daily), energizing music, and natural light to heighten the anticipation of the ultimate goal.
- Schedule moments for technological rewards at set intervals, turning waiting into a motivational resource rather than an endless drain.
- Pair routines with supportive rituals, such as lighting incense before deep concentration, stretching before rest, or initiating social check-ins to stay focused on your goals.

These subtle changes to your environment work because your brain is wired to respond to not just effort, but also the interplay of novel stimuli, contextual cues, and embedded emotional meaning. Behavioral and brain sciences research shows that thoughtful **choice shaping** (structuring your environment to make certain behaviors easier) can significantly increase the chance of completing a goal. Growth and productivity are a process of routine, creativity, and personal reward.

🔍 A CLOSER LOOK
Friction

> Friction is the resistance or ease that has been purposely included in your environment to either block or enable certain behaviors. Removing friction from good activities can mean placing your journal on your desk, your running shoes by your bed, or your water bottle at arm's reach. Adding friction to counterproductive behaviors can mean logging out of social media, silencing your phone, or placing unhealthy snacks out of sight.

By reimagining your environment through the lens of motivation, you can turn every space into an opportunity for achievement. Your surroundings don't merely support progress; they orchestrate it, making the path toward your aspirations more transparent and inviting.

🔍 A CLOSER LOOK
Choice Shaping

> Choice shaping makes desired behaviors the easiest path. When your workout clothes are laid out, your calendar protects focused time from distractions, and your workspace removes temptation, you are not relying on willpower. You are shaping the choices available so that good decisions feel inevitable rather than effortful. Your brain follows the path of least resistance, and choice shaping directs that path toward your goals.

CASE STUDY
Mei—Turning Procrastination Into Progress

Mei, a graduate student, continually found her resolve sabotaged by a familiar trap: high expectations for herself clashing with a persistent cycle of inertia and self-doubt. What began each morning as a plan to tackle her graduate thesis slipped into hours of distracted social media scrolling, reorganizing class materials, and an internal dialogue of self-criticism—meanwhile, deadlines crept nearer.

The turning point came when Mei reframed her challenge with motivation through the lens of brain science. She recognized that unhelpful cues such as messy desktops, constant notifications, and easy access to diversions weren't signs of weakness but evidence of a misaligned environment. Her neural pathways, primed for distraction, made even meaningful work seem daunting.

Together, we focused on reinventing Mei's immediate space to gently nudge her brain toward action. She created a dedicated "focus corner" by the window, furnished with a comfortable chair and only the materials she needed for the day's task. Meanwhile, she hid unrelated obligations out of sight. Digital temptations—social media, messaging apps, half-finished low-priority tasks—were placed out of immediate reach, introducing just enough friction to disrupt the habit of losing focus.

To infuse her routine with positive anticipation, Mei established a simple ritual: brewing her favorite tea exclusively for study sessions, kicking off work with a playlist of songs that evoked her sense of unlimited potential, and using a "progress jar"—adding one colorful bead to a jar for every concentrated hour of work in order to track her progress in a fun way. In this new context, her environment quietly became an ally in motivation and success, encouraging her to stick to a pattern of engagement that didn't feel like a struggle.

Within a short span, Mei noticed that starting work on her thesis, a once seemingly insurmountable hurdle, now almost happened on its own. Where she once felt overwhelmed with self-doubt, she discovered a growing sense of her own capability. As her physical and mental spaces synced, she shifted from getting stuck in anxiety to enjoying a rhythm of accomplishment, celebrating every step along the way.

Mei's journey underlines a crucial insight: When you align your surroundings with your aspirations, you alter the foundational signals guiding your motivation. Progress, rather than being a battle of self-discipline, becomes a natural product of clever design. It's a change possible for anyone willing to experiment with their world.

Beyond the Individual: How Culture and Hidden Social Forces Ignite Drive

Motivation in adulthood is rarely a solo effort. Across continents and centuries, accomplishment has depended not just on personal resolve, but also on the influence of others. Research in anthropology shows that in early societies, breakthrough ideas weren't born in isolation: They arose through collective rituals, storytelling traditions, and shared challenges that were resolved around communal fires. Monumental achievements, scientific revelations, and enduring legacies depend on collaboration, cultural structures, and mutual support.

This is in part because the anticipation and pleasure systems in the brain are most powerfully activated by affirmation, encouragement, and genuine belonging. Brain imaging reveals that dopamine's reward prediction error spikes higher when success is witnessed or celebrated within trusted circles. This explains why a lively workspace, a kitchen full of engaging friends and/or family members, or a group that debates and dreams together can make effort feel more meaningful.

Social settings can make or break motivation. When you are surrounded by people who undermine your goals, offer "feedback" through criticism, or constantly compare you to others, your dopamine system downregulates in response to that perceived social threat. Your brain registers that **social friction**—resistance, judgment, and misalignment with your values—as a threat, depleting energy and motivation. Meanwhile, small positive changes can transform your social landscape: joining a community of learners who share your aspirations, finding a trusted partner who reinforces rather than questions your path, or shifting your social media feed toward people and content that spark genuine inspiration rather than comparison. These shifts reset your neural reward system and fuel sustainable motivation.

A CLOSER LOOK
Social Friction

Social friction is the relational resistance and emotional depletion you experience in contexts where your goals are not supported. It shows up as subtle discouragement, well-meaning criticism that lands as shame, or the constant presence of people whose priorities conflict with yours. Unlike productive friction that blocks distractions, social friction blocks your forward momentum. Reducing it by choosing your social environment consciously is not selfish. It is neurologically essential for motivation to flourish.

Recent research on **group attunement** (or neural synchronization) shows teams that are highly attuned to each other can enter a zone of collective creativity, where self-awareness fades into coordination and each participant experiences new learning. Anthropologists observe that these "zones" are ancient tools of survival that allowed ancestors to gather, adapt, and thrive in sync. To activate these hidden forces, try reshaping your social surroundings by seeking out people and communities that both support and inspire you. Additionally, celebrate together: Mark milestones with shared rituals, such as collective storytelling or group reflection.

A CLOSER LOOK
Group Attunement

Group attunement is the biological state where mirror neurons activate across individuals simultaneously, creating a shared emotional and mental space. When a group achieves attunement, decision-making becomes faster, creativity expands, and individual resilience strengthens. Your brain chemistry literally shifts from stress to growth when surrounded by attuned people.

The cultures around you—built by both tradition and ingenuity— are catalysts for sustained motivations and accomplishment. By being intentional about your social environment, you can transform striving from an isolated effort into something that offers belonging, possibility, and lasting impact.

How Deeper Relationships Amplify Reward

Building on the influence of social setting and culture, the most profound catalyst for motivation is often found in personal relationships. The people you trust, collaborate with, and lean on boost purposeful living. Transformation is magnified by the encouragement and gentle accountability of friends and colleagues. It goes to show that meaningful bonds don't just support progress—they accelerate and shape it.

Contrary to common approaches that treat drive as entirely your own responsibility, modern neuroscience reveals that no aspect of your reward and motivation circuitry is more deeply shaped by authentic connection. Dopamine pathways linked to achievement respond not just to solitary victories, but to small moments of shared insight, affirming gestures, laughter with others, and collaborative effort. Brain scans show that even small gestures—an inside joke, wordless reassurance, or a thoughtful message—boost baseline levels of both dopamine and oxytocin, easing stress and enhancing resilience for hours.

Anthropology and sociology echo this truth. Throughout history, tribes, guilds, and kinship groups provided a foundation for ambition, sustenance, and risk. These social networks offered not just material advantages, but emotional validation—a force that reliably curbed feelings of isolation and protected focus and hope.

You can harness the motivating power of relationships in both in larger shifts and daily rituals by:

- Joining periodic "creation sessions," group sprints, or co-learning gatherings—even when the focus is on mundane tasks
- Offering and receiving small acknowledgments (a genuine "thank you," a note of encouragement, or shared laughter)

- Celebrating each other's progress and offering support during renewal and setbacks

By choosing to invest in and rekindle bonds that inspire, you refine your compass for purpose and supercharge motivation, mixing it with creative vigor, adaptability, and genuine joy. At their best, relationships turn solitary striving into a collective symphony—each connection tuning the following note for fulfillment.

Practical Strategies: Orchestrating Dopamine Daily

Pause and practice: Translating these chapter insights into action is how you can make long-term changes. Take this space to put the previous lessons about motivation into practice:

1. **Ignite with ritualized beginnings.** Start each day, or any significant endeavor, with a deliberate cue that makes you feel eager. This might be setting a heartfelt intention, arranging an inviting workspace, connecting with an accountability partner, or initiating a personal priming ritual. Concentration and positive expectancy both increase when "launch moments" are associated with repeated cues.
2. **Tune into your natural rhythms.** Pay attention to your fluctuations in mental clarity, inspiration, or sociability throughout each day you have started with eagerness. By timing your most meaningful efforts to these unique waves, you leverage the biology of **ultradian cycles** and enhance your creative stamina.

A CLOSER LOOK
Ultradian Cycles

Ultradian cycles are the natural 90–120-minute rhythms of high performance followed by natural dips in energy and focus that your brain experiences throughout each day.

3. **Diversify reward experiences.** In tandem with the previous step, alternate standout accomplishments (major tasks, project milestones) with subtle, replenishing pleasures, such as momentary sensory pauses (a 30-second stretch with deep breathing, a sip of your favorite tea, stepping outside for fresh air), small affirmations, or marking micro-wins. This variety maintains neural reward pathways' responsiveness and helps avert habituation.

4. **Build recovery and reflection pauses.** Plan for intervals that invite decompression—such as strolls outdoors, absorbing music, lighthearted conversations, or healthy daylight breaks. These spaces allow your nervous system to recover and integrate learning before you re-engage.

5. **Cultivate social momentum.** Deliberately infuse partnership, support, and communal celebration into your pursuit. Collaborators and shared feedback strengthen not just discipline, but also enjoyment and accountability as goals unfold.

6. **Revisit and refresh regularly.** Reserve time weekly to survey what's fueling your best energy and what feels stale. Rotate elements, experiment with new challenges, and honor adaptability—neuroplasticity thrives on reflection and iteration.

Why this matters: When you weave these practices into daily life, motivation becomes less of a resource you have to actively summon and more of an ecosystem that is intrinsic to your brain. Your **dopamine signature**, or neurochemical pattern, will evolve with you: responsive, robust, and attuned to the contours of a life in continuous creative bloom.

A CLOSER LOOK
Dopamine Signature

> Your dopamine signature is the specific way your brain responds to motivation across different contexts. You can think of it as your neural fingerprint of motivation, shaped by the choices and environments you build over time.

Navigating Setbacks and Evolving Your Motivation Blueprint: When Your Dopamine System Recalibrates

Life is shaped by more than motivation and milestones; it is also defined by the pauses, detours, and unexpected turns that challenge and reveal your abilities. Even solid routines and supportive environments can be tested by times of inertia, difficult circumstances, or the familiar pull of old habits. These moments aren't abnormal or signs of failure—they're part of personal growth.

Recent neuroscience affirms that lasting motivation doesn't come from steady uphill progress but rather cycles of progress, disruption, and renewal. The brain doesn't grow stronger just through success; it is also strengthened through the repair and recalibration that follow setbacks. When motivation wanes or enthusiasm stalls, it's an invitation to pause and observe—instead of rushing to fix things, listen to what needs rest or change before moving forward.

Throughout history, cultures have thrived by adapting with curiosity and resilience as the world shifts. The ability to face adversity with openness—seeing it as an opportunity for learning rather than a threat to your identity—is the hallmark of true endurance. Communities and individuals alike thrive when they honor the natural ebbs and flows of life rather than resisting them.

Your unique motivation blueprint comes in moments of triumph as well as in reflection, redirection, and the courage to start again. Rather than striving to be perfect, learn to work with life's natural fluctuations; each cycle adds valuable experience and a foundation for future adaptability. When disappointment or a sudden dip in drive appears, recognize it as part of authentic progress. Each return builds deeper motivation, and each adjustment adds insight. True achievement includes effort, contemplation, renewal, and reconnection—celebrating both the peaks and the valleys that shape a life well lived.

CHAPTER SUMMARY

Cultivating sustainable drive means moving beyond willpower and quick fixes: Embrace the brain's cycles of anticipation, accomplishment, connection, and rejuvenation. Every environment, relationship, and ritual can be shaped to support this inner orchestra, turning daily experience into continued motivation. Keep these takeaways from this chapter in mind as you move on to the next:

- True fulfillment comes not from striving for constant excitement or flawless routines, but by harmonizing action, renewal, and reflection in service of authentic goals.
- Dopamine becomes a reliable ally when you link it to rich environments, supportive relationships, and rituals that feed curiosity and satisfaction.
- The spaces and cultures you inhabit can magnify or mute motivation—intentionally shaping them boosts your success.
- Lulls and setbacks are integral phases of any growth journey; stay driven by responding to these situations with openness, adaptability, and self-compassion.

CHAPTER 12

The Dopamine-Driven Workplace

Finding Focus and Flow at Work

Modern work promises countless opportunities for connection, growth, and achievement, but it often delivers an exhausting blend of overstimulation, pressure, and uneasy striving. Today, from cubicles to home offices, the very systems designed to maximize productivity fuel distraction, self-doubt, and a chronic sense of not quite measuring up. Beneath the surface of project deadlines, strategy meetings, and digital dashboards lies a deeper story that involves dopamine.

In small doses, dopamine sparks attention and innovation and drives routine taskwork. Yet the relentless novelty and reward cycles engineered into modern employment—from urgent emails to constantly changing goals—have many stuck in a loop of anxious overdrive and paralyzing depletion. What's meant to keep you focused can easily push you into restlessness, comparison traps, and the belief that everyone else is coping better. Meaningful work is more than output or status—it is a source of belonging, identity, and creative contribution. But when insecurity or imposter syndrome take over, the workplace becomes a space of unmet expectations, harsh self-criticism, and stress. Evolution equipped humans with the ability to thrive in collaborative, rhythmic environments involving effort *and* renewal—not the perpetual hustle or remote isolation that so often defines contemporary professional life.

This chapter reveals how a dopamine-driven approach can transform your relationship with work, shifting endless busyness into deliberate engagement. Through neuroscience and psychology, you'll discover practical strategies for restoring focus, fostering resilience, and rekindling genuine satisfaction, regardless of your job title or industry. And you'll meet Rachel, who learned to navigate the dopamine traps of the modern workplace. Here, the science of motivation becomes a tool kit for not just *surviving* another workday, but *thriving* with confidence, insight, and purpose.

The Neuroscience of Modern Work: Why You're Wired for Both Stimulation and Exhaustion

The world of work has changed dramatically—both in what is required of people and in how the brain responds to these shifts. Open-plan offices and remote teams alike are built for novelty and urgency: endless notifications, shifting priorities, and the expectation that productivity means perpetual motion. Every message ping or task-completed alert delivers a small hit of dopamine.

In moderation, this response is adaptive. Dopamine evolved to help your ancestors focus on what mattered, solve new challenges, and feel motivated when navigating uncertain environments. In the workplace, that same chemistry can infuse the start of a new project or the setting of a big goal with genuine anticipation and drive. Early stages—learning the ropes, launching a campaign, solving a problem—often deliver a motivating cocktail of excitement and purposeful energy. Yet problems begin to emerge as the stimulation grows. Repeated deadlines, alerts, praise, and criticism can desensitize the dopamine system. Tasks that once felt fresh now require greater effort for the same level of satisfaction, while distractions multiply and recovery feels impossible. Psychologically, this leads to a paradoxical fatigue: Your mind races while focus fades, and the most talented, hardworking individuals find themselves running on fumes before the end of the workday.

As you explored earlier in this book, human beings are designed for cycles of effort and rest—for work that feels meaningful and communities that provide feedback and support. Modern workplaces often hijack these ancient systems, making it difficult to focus, stay motivated, or know when to truly switch off your work brain. When self-worth becomes tied to email inboxes, coworker comparisons, and constant performance, fatigue becomes less as a signal than a permanent state of being.

Understanding this mismatch between biology and environment is the first step to curbing distraction and a **burnout spiral.**

A CLOSER LOOK
The Burnout Spiral

Typical workplace burnout leaves you feeling emotionally drained and disconnected from your job. Your energy and emotions are depleted, you may develop cynical feelings toward your work, and your professional effectiveness suffers. Over time, burnout builds gradually into a burnout spiral, where you become stuck cycling between constant demands for high-output performance and inadequate recovery.

When you see fatigue and lack of focus not as failures of willpower, but as natural reactions to modern pressures, you can start to find—and design—new rhythms that honor both the brain and the spirit.

The Distraction Dilemma: Why Focusing at Work Is Harder Than Ever

Today, attention has become both a prized asset and a challenge. Never before have workers had access to so many tools for connection, collaboration, and information. Many find themselves losing hours to shallow tasks, digital alerts, and mental drifts. On the surface, it appears to be a problem of willpower or discipline. However, this isn't simply about self-control; it's about how your attention and reward systems interact.

Dopamine drives you to complete tasks. Each time you tick off an urgent item on your to-do list, respond to a message, or chase down a new lead, your reward system gives you a slight boost that feels both productive and satisfying. But when the stimulation never stops, your attention is constantly pulled from one novelty to the next, making it harder to focus deeply on your goals.

Research shows that while the mind can rapidly shift focus, multi-tasking is essentially a myth. Switching focus drains both mental energy and dopamine, leaving you more tired, less fulfilled, and more prone to error. You have more difficulty sinking into "the zone" where your best work gets done. Evolution wired the brain to scan for threats and opportunities, not manage dozens of competing tasks. Workplaces that prioritize busyness over depth or reward constant responsiveness instead of intentional progress only deepen the imbalance.

Recognizing that lapses in focus is not a personal failure but an outcome of modern brain–environment mismatch is freeing. It allows you to approach focus as a trainable skill, rooted in both biology and behavior, rather than a fixed trait you either have or don't. Later in this chapter you will explore concrete, science-backed strategies for reclaiming attention at work—transforming productivity from a source of exhaustion into a pathway for meaningful contribution and genuine satisfaction.

CASE STUDY
Rachel—Battling Insecurity and Restoring Focus

Rachel, a junior project manager at a busy bank, had always taken pride in her hard work and willingness to go the extra mile. But her days had become an exhausting cycle of self-doubt, anxiety, and feeling perpetually behind on tasks. Each morning, she would open her overflowing inbox, already feeling tense and believing that her colleagues were silently questioning her competence. Despite her best efforts, she just felt more distracted and anxious—jumping from report to report, answering every "urgent" request, and struggling to finish anything meaningful.

Underneath her relentless pace was a silent battle with imposter syndrome—an inner voice that insisted every mistake or delay was "proof" she wasn't cut out for her role. This anxiety impacted her relationships at work: She often hesitated to ask for help, feeling it would be seen as a sign of incompetence. Instead, she held herself to ever-higher standards and projected her worries onto her team, becoming very critical or withdrawn when her stress peaked. This only seemed to deepen her sense of isolation and exhaustion. Meetings became draining rather than connecting. The satisfaction she had once felt in learning a new system or delivering a successful project was replaced by short-lived relief, quickly overshadowed by new pressures and self-critique. Rachel's brain, wired to scan for mistakes and anticipate the next crisis, rarely allowed her the dopamine boost that comes from enjoying progress or celebrating wins.

When Rachel reached out to me for coaching, she wasn't seeking a productivity hack—she wanted to reconnect with the purpose that had once made her work feel meaningful. Our conversations began by dismantling the myth that her struggle was a sign of weakness. Together, we mapped the invisible pressures of the modern workplace onto her reward system. She learned to recognize the telltale signs of dopamine depletion: the restless need for novelty, the inability to unwind, and emotional numbness.

Rachel then started to embrace tiny, actionable shifts. She experimented with clear boundaries around digital distractions and prioritized completing tasks over reacting to every new one. She also built regular pauses for reflection and feedback into her daily routine. Over time, she became more confident and satisfied—not by working harder, but by partnering with her brain rather than pushing against it. Her journey shows that resilience and fulfillment at work depend not just on effort, but also on understanding, honoring, and retuning the motivational circuits within.

The Professional Brain: Strategies for Reclaiming Motivation at Work

Motivation and purpose at work can be found through a few reinforcing factors. From flow states to social connection, the brain can be aided in engaging, persisting, and thriving. Understanding these distinct strategies will help clarify your own approach to the workplace and identify concrete paths to better engagement on the job.

The Flow State

True professional satisfaction often pops up in those situations where time slips away and work feels nearly effortless—moments of full absorption that psychologists call "flow." Achieving flow on the job isn't about having the perfect role or working in a distraction-free bubble; it's about partnering with your brain's natural motivation systems to create the right conditions for focus, purpose, and achievement.

Neuroscience shows that flow is a coordinated dance of multiple brain networks, with dopamine playing the starring role. When your skills match the level of challenge—not too easy, not too overwhelming—the brain responds with a focused release of dopamine that helps you concentrate and stay motivated. Distractions fade; instead of juggling other tasks or battling anxiety, you experience a "sweet spot" of engagement—you feel energized but not frantic, curious but not scattered.

Evolution shaped the flow state, and for good reason. In your ancestral past, performing a task well (tracking subtle patterns, solving complex problems, collaborating under pressure) often meant the difference between life and death. Today, while the dangers are usually less extreme, the same brain systems help meet the demands of work. Brain scans show that during flow, not only does the prefrontal cortex (responsible for planning and self-monitoring) operate more efficiently, but key reward centers light up, reinforcing the desire to return to deep, meaningful work again and again.

However, modern workplaces make it hard to enter a true state of flow. The barrage of notifications, conflicting priorities, and pressure for constant output block deep concentration before it can even start. The very design of many workplace cultures actively works against flow: goals are unclear, feedback is inconsistent, and multitasking is an expected norm. The result? Even well-intentioned professionals find flow challenging and focus impossible. It's that problem of brain–environment mismatch showing up when you most want (or need) to get things done.

Fortunately, by understanding the science of flow, you can create the conditions for it. Clarifying goals, structuring tasks to a manageable level of difficulty, seeking regular feedback, and protecting blocks of uninterrupted time will help you tap into a state of flow. You'll be able to practice getting into a flow state in the "Practical Strategies" activity in this chapter.

Social Connection

Social connection stands out as a particularly powerful strategy for reclaiming drive in the workplace. Behind every spreadsheet, strategy session, and solo project is a deeper need to connect with others. Neuroscientists have discovered that the urge to belong—to share effort, ideas, and emotion—activates reward and motivation circuits in the brain just as powerfully as personal achievement. In other words, the brain is wired to find motivation not just in individual success, but in working and growing with others. Evolutionarily, the workplace is a modern extension of the tribes and bands where your ancestors' survival depended on collaboration, mentorship, and trust.

When you solve problems with others, receive encouragement, or engage in empathetic conversation, dopamine and other chemicals in the reward system are released. These moments nourish the areas of the brain where emotional strength and pride, flexible thinking, and long-term satisfaction come from. In fact, studies (e.g., Google's "Project Aristotle" study of 2015) show that workplaces with a culture of authentic support—where team members regularly express appreciation, share credit, and solicit honest feedback—consistently outperform those that prioritize only competition or solo productivity.

However, many modern organizations unknowingly cut off social rewards. Virtual meetings may fill the calendar, but they typically lack the spontaneous humor or nonverbal cues that reaffirm trust in a face-to-face situation. Workplace surveys increasingly report that even the most talented employees struggle to feel engaged or motivated when interaction feels transactional or superficial. Neurobiologically, this can lead to what psychologists call **social pain**: a state of depleted dopamine, blunted motivation, and disconnection from purpose.

🔍 A CLOSER LOOK
Social Pain

Social pain activates the brain regions responsible for processing threat and loss. When someone feels excluded, dismissed, or forced into transactional interactions, their nervous system interprets this as a social threat. The result is less motivation, flattened mood, and a sense of purposelessness that persists even in high-performing environments. In workplaces, social pain manifests as disengagement despite talent or opportunity. Remote work, fragmented communication, and efficiency-focused meetings intensify social pain by eliminating the incidental moments of genuine connection that recharge dopamine and reinforce belonging.

The good news is that change does not require an overhaul of workplace culture. Small acts—such as starting meetings with a personal check-in, sharing a recent challenge or success, or expressing appreciation for a colleague—make a big difference. These rituals of connection serve as dopamine boosters, helping professional teams weather pressure, navigate setbacks, and celebrate progress, even amid high demands or uncertainty.

When workplaces cultivate spaces for laughter, vulnerability, and shared wins, they do more than boost morale—they activate the deepest circuits of engagement, learning, and psychological safety in the brain. No matter where you may work, remember that it is fundamentally social. This insight will help you turn the pursuit of productivity into a more sustainable, humane, and ultimately rewarding journey.

Motivation Rituals: Designing Your Own Dopamine-Driven Practices

While lasting motivation and focus are built in the brain, they also need to be supported in everyday life. The difference between enduring satisfaction and chronic frustration at work rarely comes from big gestures or major leaps, but from small, intentional shifts—simple, regular

The brain is wired to find
motivation not just in
individual success, but
in working and growing
with others.

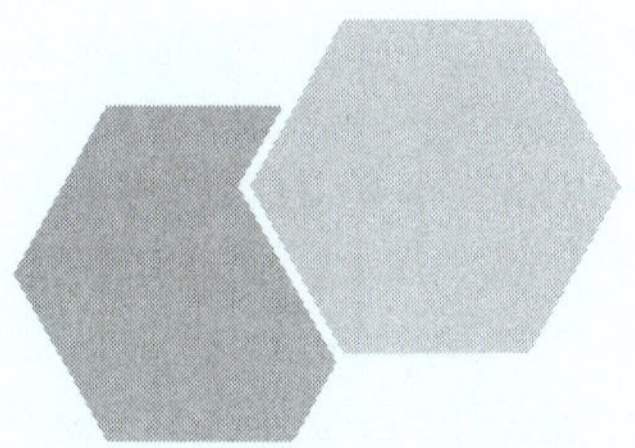

rituals that gently boost dopamine, sharpen your mental focus, and spark meaningful involvement in your work.

Starting your workday with specific cues primes your attention for what matters. This might be as simple as reviewing a handwritten priority list before opening your inbox, lighting a scented candle before making a challenging phone call, or taking a moment to stand by a window to clear your mind. These rituals signal to your brain that it's time to shift from reacting to taking charge.

Science shows that consistent routines help the brain build shortcuts, making it easier to get into a focused and motivated state when distractions pop up. Ending the workday with simple closing rituals is just as important. These may include shutting down your computer, stretching, writing down one thing you learned or accomplished, or thanking a colleague. These small acts reinforce a sense of progress and control, helping dopamine circuits recognize both effort and completion. Where possible, you can also bring others into your motivation-building rituals: Brief morning huddles, weekly "wins" discussions, or quick feedback can turn fleeting inspiration into shared momentum, resilience, and pride.

Importantly, revisiting and adjusting your rituals as your work needs change can further help you navigate uncertainty and bring renewed purpose to every workday. What sparks focus during a busy week may not serve you during a period of recovery; the ritual that grounds you in new responsibilities might evolve as your confidence in your ability grows. Treating these rituals as dynamic tools rather than rigid routines fosters productivity, a sense of belonging, and mastery—both for your professional team and for yourself.

Practical Strategies: Boosting Focus, Motivation, and Satisfaction at Work

Pause and practice: Now that you have learned about the challenges and strategies for staying motivated at work, take about four weeks to experiment with these practices. Harnessing the brain's motivational chemistry can transform your workday from a cycle of stress and distraction into one of progress, engagement, and renewal:

1. **Start with a dopamine primer.** Begin the day with an energizing ritual—a few minutes of sunlight, brisk movement, or tackling a favorite task. This sends a signal to your brain that it's time to shift into purposeful focus, offering an early, positive dopamine boost.

2. **Clarify challenges and goals.** Break larger projects into short, well-defined milestones. The sense of progress and closure that comes when you complete each step activates your brain's reward circuits, promoting lasting motivation.

3. **Guard your golden hours.** Identify and protect the times you naturally feel most alert or creative. Block interruptions and dedicate these windows to meaningful work. Leave admin tasks or emails for lower-energy periods.

4. **Design micro-breaks.** Integrate regular, active breaks between tasks. Stretch, drink water, or step outside. These recovery moments help restore your dopamine balance, combat decision fatigue, and maintain cognitive resources for deep thinking.

5. **Connect and collaborate.** Make space for brief check-ins or creative huddles with supportive colleagues. Social connection, especially when paired with shared problem-solving, releases powerful reward signals in the brain—and also offers new perspectives.

6. **Celebrate wins, however small.** Take a moment to acknowledge completed tasks or team achievements—whether by writing down successes, giving public praise, or sharing gratitude. Savoring each milestone reinforces positive motivation and makes you more prepared for the challenges ahead.

7. **Reflect and recalibrate.** At the end of each week, scan your routines to identify what acts lifted your focus, what drained it, and what could be tweaked. Treat this as an ongoing experimentation—an act of self-leadership and self-compassion, not self-critique. After two weeks, step back and look for patterns across your weekly check-ins. Identify which practices are genuinely moving

your motivation baseline and which ones still feel forced. Then, at the end of the month, do a deeper review. Ask yourself what shifts you've noticed in dopamine stability, motivation quality, and engagement. Decide whether to deepen a practice, swap something out, or introduce a new element. Also consider any seasonal changes or work cycle shifts that may call for adjustments.

Why this matters: There's no rigid template for thriving at work. Still, by tuning into your brain's rhythms and building intentional, dopamine-driven habits, you can foster productivity, sustained energy, and profound satisfaction—no matter your role or work environment.

CHAPTER SUMMARY

Work can be either a place for your brain to thrive or struggle with dopamine imbalance. Keep the following points from this chapter in mind regarding dopamine and your workplace:

- Dopamine serves as the brain's signal for engagement and progress at work, but unrelenting stimulation and productivity pressure can erode motivation, satisfaction, and well-being.
- Constant distraction isn't a personal failing, but a mismatch between modern work demands and ancient brain systems.
- Burnout emerges when workplace cultures prioritize output over recovery, undermining both productivity and the neural circuits that drive creative problem-solving.
- Lasting motivation can be found in designing daily rituals—anchored by clear goals, restorative breaks, and authentic connections—that gently tune your dopamine system for sustained attention and deep fulfillment.
- By understanding and working with your brain's natural cycles, you can transform your workday from mere survival into a source of energy, mastery, and meaning—for yourself, your team, and your larger purpose.

CHAPTER 13

Dopamine and Creativity

Harnessing the Power of Novelty

Creativity isn't reserved for a gifted few; it's an inherent human ability that is built into the brain. As you uncovered earlier in this book, everyone is born with a drive for novelty, surprise, and meaningful challenge. It's a built-in craving that has fueled innovation across cultures and throughout history. At the center of this inventive spirit is dopamine, which is highly sensitive to the allure of new possibilities. It powers your "seeking system," motivating exploration, storytelling, creative breakthroughs, and the urge to reinvent the status quo. But creativity brings more than fleeting excitement. It restores energy, gives purpose, and offers a sense of belonging to something greater than daily routines.

However, the same forces that energize new ideas—exploration, daring, bold experimentation—can overwhelm the brain and its creative systems. When dopamine levels drop, inventiveness stagnates; conversely, when too much dopamine floods the system, enthusiasm can turn into anxiety, sleep loss, or burnout. Research shows that sustained creativity requires not just boldness and risk but also balance, recovery, and self-awareness. This chapter takes a closer look at creativity and novelty, from the science to how you can harness ingenuity in your Dopamine Menu. Here, you'll explore rituals, insights, and environments that consciously support creativity—turning novelty into a steady source of personal and communal progress. Using a holistic approach backed by neuroscience, you will learn how to turn your own brand of creativity into a lasting force for fulfillment rather than a random spark.

Modern life inundates you with stimulation, endless rewards, and boundless opportunities. While fresh ideas are easier to pursue, overload and exhaustion are also easy to fall into. This chapter will also help you build a dopamine tool kit for embracing your creativity. You'll practice the tools of resilience, flexibility, and emotional breadth while safeguarding overall well-being. By working with the natural cycles of the mind and body, you'll add another helpful layer to your Dopamine Menu.

Creative Drive: Dopamine, Novelty, and the Roots of Innovation

The urge to invent and experiment is deeply rooted in your biology. Your brain is primed for exploration, given it is powered by dopamine, the chemical messenger that fuels motivation, learning, and discovery. Dopamine acts like a compass, guiding you toward new obstacles, unexpected experiences, and the chance to make something remarkable from the ordinary. Neuroscience reveals that this drive for discovery involves both rapid and sustained releases of dopamine. Bursts of dopamine occur when you find new solutions, while steadier levels of dopamine help you focus, reflect, and turn creative concepts into finished work. Studies affirm these dual roles: When you step outside your routine, dopamine-rich areas of the brain, such as the striatum (particularly the ventral striatum) and prefrontal cortex, become activated, shifting you from old habits to flexible, creative thinking.

Creativity is also shaped by culture and community. From storytelling and music to tools and art, the hunger for the new shapes society. Over time, these communal practices strengthen not just relationships themselves, but collective wisdom and innovation. Environments that celebrate curiosity, encourage safe risk-taking, and nurture openness help creativity flourish; meanwhile, those that favor conformity or a fear of new things tend to suppress it.

Recognizing both the biological origins and cultural contexts of creativity allows you to take a more intentional approach to using this energy. True creative mastery comes from a balance of adventurous exploration with thoughtful reflection. Sustained growth happens when you follow quick bursts of inspiration with mindful effort.

Creative Risk-Taking: The Rewards *and* Realities

Every creative act involves taking risks—whether launching a business, sharing artwork, or proposing a new solution. Advancement requires stepping outside comfort and routine. Dopamine drives the excitement of these healthy risks, pushing you to explore new challenges and embrace uncertainty. (This natural response is thought to be partly due to the fact that evolutionarily, people who explored and adapted had a much better chance of not just surviving, but thriving.)

But every leap into innovation also makes you vulnerable. In each risk, there is a possible rejection, failure, or criticism. This triggers the brain's threat response, tempering dopamine-fueled enthusiasm with caution. Managed well, this tension between excitement and fear sharpens focus and channels nervous energy into preparation—transforming apprehension into the fuel for progress. When one force dominates and suppresses the other, the outcome becomes counterproductive. If enthusiasm runs unchecked and caution is ignored, you move forward recklessly, without adequately preparing for or assessing the risks. This often results in preventable mistakes, poor strategic decisions, and unnecessary losses. Conversely, when caution takes control and enthusiasm is suppressed, you become paralyzed by worst-case scenarios. You may abandon promising ideas before testing them, avoid necessary growth opportunities, or retreat into the safety of what's already known.

Environments that punish mistakes or instill fear of judgement can dramatically inhibit the kind of healthy risk-taking that leads to innovation. Historically, major advancements in culture, science, and art have flourished in societies that embrace experimentation, encourage learning from missteps, and provide room for growth. Acts such as public storytelling, group improvisation, or reframing setbacks as lessons foster resilient, adaptive approaches to innovation and help build confidence.

For today's changemakers, welcoming risk means embracing discomfort as a natural part of growth and understanding that personal and creative development come from uncertainty. Trying a new artistic medium, proposing unconventional projects, or sharing a unique perspective requires courage. Pairing risks with self-reflection, support networks, and periods of recovery turns uncertainty into authentic achievement and an enduring creativity.

Burnout: When Dopamine's Creative Fire Backfires

For many creative minds, the intense rush of innovation can quickly morph into feelings of fatigue, emptiness, or overwhelm. This exhaustion is typical in cultures that prioritize achievement and is often perceived as inevitable. While dopamine fuels creative exploration and growth, excessively stimulating its circuits and leaving no space for rest can lead to a lack of motivation and diminished enthusiasm.

Research indicates that peak periods of creativity boost dopamine, which invigorates your reward centers and keeps you motivated. However, the brain quickly adapts to ongoing stimulation. As novelty or progress fades, so does drive. Eventually, more and more effort is required to feel truly engaged in an activity. You become less tolerant to ambiguity or setbacks and may ultimately feel irritable, lethargic, and emotionally detached.

Burnout rarely comes from overwork alone, however. Underlying factors—such as the pressure for perfectionism, the stress of unpredictability, social isolation, or a lack of rest—erode motivation over time. Persistent demands, consistent digital feedback, or harsh self-criticism can overwhelm the prefrontal cortex and diminish mental flexibility and enjoyment, making routine responsibilities feel impossible.

For your ancestors, stepping back when confronted by relentless strain was key to survival. It signaled a time to recover energy and reevaluate what pursuits were worth working toward.

Recognizing that cycles of innovation *and* recovery are an ancient part of your biology—not a personal failing—can bring relief. Lasting creativity requires boundaries, rest, and respect for the natural fluctuations of your mental and physical energy. Later in this chapter, you will find actionable strategies, restorative habits, and social supports to balance your dopamine systems and curb burnout.

The Creative Advantage: Building a Dopamine Tool Kit

Lasting creativity rarely comes from nonstop effort or fleeting epiphanies. Those who consistently advance new ideas know that sustaining innovation requires self-awareness and an ongoing, intentional approach. Research continues to show that originality thrives when creators keep track of their mental rhythms—mindfully noting shifts not only in enthusiasm and fatigue, but also in curiosity, energy, and focus.

Dopamine is a resource, not a bottomless well. Those who thrive treat it like a finely tuned engine, balancing productive work with rest and playful exploration. This rhythm is often the hidden advantage behind breakthroughs. Whether it's a design team finding renewal in nature, an artist pausing for a fresh perspective, or a scientist recharging through collaborative play, breakthroughs often arrive unexpectedly after the mind has a chance to rest and reset ideas.

A dopamine tool kit is a collection of practices and environmental supports that help you manage your creative energy. Rather than treating your dopamine system as a resource to exploit endlessly, this tool kit provides concrete strategies for sustaining innovation without burnout. The primary tools of a dopamine tool kit include:

- **Intentional novelty:** Unexpected elements, fresh environments, or unfamiliar mediums that prime your dopamine system and reawaken curiosity
- **Balanced pace:** Alternating cycles of intense creative exploration with deliberate periods of refinement, reflection, and rest to prevent innovation fatigue

- **Creative bursts:** Focused intervals designed with recovery time built in to allow ideas to grow and protect the mind against overload
- **Celebration of completion:** Markers at the end of meaningful work, such as reflection or shared recognition, that reinforce accomplishment and sustains motivation
- **Social connection:** Regular collaboration, a sharing of ideas, and constructive feedback that allow creativity to flourish collectively
- **Weekly reflection:** Assessments of which practices elevated your energy and which ones drained it, providing insight that can help you adapt to your rituals as needed

Together, these elements encourage breakthrough ideas and creative stamina, transforming originality from a random happening into a lasting force for fulfillment. A well-designed dopamine tool kit reframes boredom as an invitation for new growth, treats setbacks as valuable pivots, and honors the role of recuperation in fueling future breakthroughs. Instead of draining inspiration through constant motion, innovative minds can cultivate curiosity, moderation, and a mindset for enduring possibility—turning creative fulfillment into an ongoing, renewable resource.

CASE STUDY
Alex—Riding the Edge of Innovation

Alex, a game designer, was recognized for his groundbreaking work in the gaming industry. Beneath the accolades and brainstorming sessions, however, life felt like a loop of perpetual urgency, adrenaline-powered sprints, and an overwhelming struggle to unwind after work. As demands mounted, the nights and mornings blurred together, his workspaces overflowed with half-finished sketches, and every milestone just made him more focused on the next burst of inspiration.

He described his creative process as exhilarating yet draining: "When inspiration hits, the rest of the world falls away—it's pure flow." Alex found himself relentlessly chasing that feeling; he continually escalated goals, added more product features, and tightened deadlines, as he was convinced that his creative edge depended on constant intensity. The initial waves of reward and satisfaction faded rapidly and were replaced by agitation, mounting self-doubt, and a growing sense of falling behind.

Fatigue began invading all aspects of Alex's daily existence. Relationships strained, restful sleep grew rare, and projects that once thrilled him now made him anxious. "Some days I'd power through for hours; other days I'd just stare blankly at the screen, unable to begin," he admitted. Feedback that once energized him now felt overwhelming, and the previously enjoyable hustle of a creative job began eroding his motivation.

In our coaching sessions, we shifted the focus from restraining his creativity to honoring its natural rhythms and learning how to sustain it. With a tool kit grounded in neuroscience, Alex began to recognize the subtler signs of dopamine depletion and burnout. He learned to distinguish between healthy excitement and unproductive overdrive, and he recognized when it was time to pause and recuperate. Practicing unplugged breaks, rekindling old hobbies, seeking laughter beyond work, and reconnecting with friends all restored vitality and perspective to his daily life.

Gradually, Alex cultivated greater self-awareness and respect for his personal creative cycles. He began alternating concentrated sprints with periods of reflection and restorative downtime. He even discovered that his most innovative concepts emerged during moments of calm and relaxation rather than frenzied effort. This transformation highlights a central lesson of dopamine and creativity: Sustained innovation depends on balancing drive with mindful rest—transforming burnout into renewal and deeper satisfaction.

Dark Side of Novelty: Navigating Innovation Fatigue

New ideas are thrilling. Small acts (such as starting a new project, experimenting with a creative twist, or learning a unique skill) can infuse daily life with energy and purpose. Yet beneath the thrill of innovation, constant pursuit of originality can turn joyful exploration into an exhausting race with no finish line. Many creative people end up feeling both energized *and* drained by the things that once inspired them—a kind of **innovation fatigue**. Unlike burnout, which comes from prolonged overload and depletion, innovation fatigue is an exhaustion that arises when constant novelty and rapid pivots overload your reward system, making it harder to commit, follow through, and feel satisfied with any one path.

Central to the idea of innovation fatigue is dopamine, the brain's messenger for surprise and anticipation. Research has shown that dopamine rewards you most powerfully in moments of unpredictability, reinforcing your drive to seek out and engage with unexpected opportunities. This ancient system evolved to thrive on sporadic novelty—a rare discovery, an unlikely solution, or an unanticipated burst of insight. However, contemporary work cultures and digital environments can inundate us with constant stimuli, overwhelming our ability to process and appreciate genuinely fresh experiences.

Over time, such overstimulation begins to reshape our expectations and sense of fulfillment. Achievements bring only brief satisfaction before curiosity is swiftly redirected to new pursuits. Projects that once felt exciting can quickly lose their luster if they lack surprise. Psychologists call this phenomenon the **arrival fallacy**—the fleeting joy of accomplishment diminishing almost immediately as the mind seeks its next target.

A CLOSER LOOK
The Arrival Fallacy

Despite achieving what you thought would bring lasting fulfillment, the arrival fallacy has you finding yourself no happier than before—instead, you simply seek the next goal. This is driven by dopamine's "wanting" system: The brain is wired to anticipate reward more powerfully than to savor it once achieved.

In daily life, this pattern may manifest as persistently shifting interests, frequent changes in direction, or an inability to persist through challenges. For some, it results in abandoning creative work midstream, never savoring the completion or integration of new skills. Socially, the pressure to continually keep up with trends or measure success against viral exemplars can erode both confidence and intrinsic motivation.

Historically, wisdom traditions, such as Stoicism, contemplative Buddhism, and Renaissance artistic mentorship recognized the importance of slow periods, reflection, and ritual repetition. These practices offered a counterbalance to the chaotic pace of novelty, encouraging creators to pause, consolidate, and reflect on their achievements. Without intentional integration, constant novelty can drive our motivational circuits into diminishing returns, making it harder to attain sustained joy and progress.

Ultimately, practical neuroscience teaches that enduring creativity is a process of balance. Like tending a fire, innovation requires steady fueling but also intervals of rest, consolidation, and humble reflection. By consciously pacing new projects, revisiting ongoing work with renewed perspective, and actively celebrating meaningful completions (not just new beginnings), creative individuals can transform innovation fatigue into deep, lasting satisfaction.

Practical Strategies: Activating Creativity with Your Dopamine Tool Kit

Pause and practice: As you've explored in this chapter, cultivating creativity is not merely a matter of waiting for inspiration—it's about structuring your environment, habits, and mindset to ignite ingenuity and maintain momentum over time. By applying neuroscience-informed practices, you can train your brain to strike a balance between novelty and rest, excitement and recovery, and boldness and reflection. Use these steps to build your own dopamine tool kit now:

1. **Add intentional novelty.** Before diving into creative tasks, introduce an unexpected element—a change of setting, a playful

challenge, or an unfamiliar medium. This kind of variety refreshes mental patterns, primes your dopamine system, and reawakens curiosity, opening you up to new perspectives.

2. **Find a balanced pace.** Avoid innovation fatigue by alternating the thrill of new beginnings with periods dedicated to refinement and completion. Recognize the value in both brainstorming fresh concepts *and* deepening existing ones so you foster thoughtful progress and genuine accomplishment.

3. **Try creative bursts.** Break large projects into short, focused intervals. After each burst of concentrated effort, deliberately step back and allow space for ideas to settle. This practice helps you sustain energy, prevents overload, and fosters clarity in every stage of your work.

4. **Celebrate completion.** Make an event of finishing a task or project. Whether through private rituals, sharing wins with trusted colleagues, or taking moments to reflect on your journey, honor completion to boost pride and maintain strong motivation for future initiatives.

5. **Use group engagement.** Creativity flourishes in supportive communities. Invite collaboration, share evolving ideas with others, and seek out regular feedback. Social interaction adds dimension to your efforts and transforms solitary pursuits into shared victories.

6. **Do a weekly reflection.** Take time each week to evaluate where your creative energy peaked, what challenges emerged, and which habits most effectively supported your process. Use these insights to adjust the tools in your dopamine tool kit.

Why this matters: Remember, your dopamine tool kit is not static—it should evolve continuously along with your projects, aspirations, and personal needs. By mastering these adaptive strategies, you create the conditions for breakthrough ideas, resilient energy, and long-term fulfillment, adventuring on a creative journey that feels both enriching *and* sustainable.

CHAPTER SUMMARY

Creative fulfillment is built not just on moments of inspiration, but by learning how to balance novelty, rest, reflection, and meaningful connection—transforming the spark of originality into a lasting and enriching journey. Keep these important insights in mind as you move on to the next chapters:

- Dopamine drives curiosity, adventure, and innovation, but also needs to be managed to avoid cycles of exhaustion or repetitive habits.
- Resilience allows you to anticipate and respond to innovation fatigue by blending periods of exploration with intentional rest, deep reflection, and meaningful integration.
- A personalized dopamine tool kit draws on brain science, practical rituals, and real-life experiences to encourage creative collaboration, adaptability, and lasting inspiration.
- Effective tools in a dopamine tool kit include adjusting your mindset, pacing new challenges, celebrating real progress, and reflecting honestly on what is and isn't working.

CHAPTER 14

Life after Trauma

How Brain-Based Strategies Can Restore Healthy Dopamine Function

Resilience after trauma is not simply about trying harder; it's about fundamentally changing how the brain processes fear and hope. Central to this transformation is dopamine. When trauma throws dopamine off-balance, emotions suffer, curiosity and problem-solving diminish, and daily routines become colored by vigilance, retreat, and exhaustion. As you learned in Chapter 7, neuroscience shows these changes aren't personal failures but rather the result of neuroplasticity: the brain's ability to constantly learn and rewire itself over time. Through neuroplasticity, repeated patterns of worry, fear, or avoidance become reinforced habits and belief systems. But this same neuroplasticity can be used to heal. By mindfully focusing attention and recalling even brief moments of safety, care, or empowerment, you can cultivate new neural pathways that support recovery.

This chapter explores how the science of neuroplasticity and dopamine offers trauma survivors practical, research-backed strategies to build resilience—one intentional choice at a time. By drawing from a healing array of tools ranging from focused self-awareness and movement, to safe connections and positive experiences, anyone dealing with trauma can experiment, adapt, and select the approaches that most powerfully support their recovery. In this way, recovery is as individual as each person's history; it is an active process of nurturing self-belief and lasting growth.

With guidance, the brain can learn to sort the present reality from old alarms, boosting dopamine signals that reinforce a genuine sense of safety and setting new routines that support engagement and satisfaction. Positive states of mind—nurtured through intentional awareness, movement, or recalling moments of kindness—act as seeds for new, resilient brain networks. This chapter is not about erasing pain or denying difficult emotions but about ensuring that past losses and fears do not control the narrative of life. It's about fostering the conditions for growth, curiosity, and connection after hardship.

Neuroplasticity and Dopamine: Digging Deeper Into the Science of Trauma

Neuroscience reveals that the brain is always changing; it is shaped, moment by moment, by the experiences you have—especially repeated ones. When trauma becomes a focus and conditions the nervous system to expect danger, responses of avoidance or withdrawal become automatic. As you learned in Chapter 7, dopamine plays a key role here: It helps determine which situations are worth approaching and which ones are better avoided. In the aftermath of trauma, this system can become skewed: Dopamine pathways reinforce avoidance even in situations that aren't actually threatening, because each time the person escapes discomfort—even temporarily—they receive a subtle neurochemical "reward."

This is known as **stress-induced neuroplasticity**. Through this process, the mind becomes focused on predicting danger, recalling losses, and bracing for disappointment. Given this stress response, it's no wonder that, for many trauma survivors, basic joys and the excitement of possibilities dull.

🔍 A CLOSER LOOK
Stress-Induced Neuroplasticity

> Stress-induced neuroplasticity is the brain's capacity to physically restructure neural circuits in response to stressful experiences. This process involves strengthening synapses and reactivating certain neural networks.

The good news is that neuroplasticity also enables healing. The very circuits that wire fear and avoidance into your brain can be redirected toward growth with careful, consistent practice. In the next sections of this chapter, you will uncover how to use the science of neuroplasticity to rewire the brain for healing.

Pathways to Healing: Four Dopamine-Based Strategies for Rewiring Resilience

Healing from trauma rarely follows a single path. The brain's remarkable capacity for change means it responds to multiple types of input, from structured therapeutic interventions to spontaneous moments of unexpected comfort. The four strategies that follow are supported by neuroscience research and clinical practice. The goal isn't to master all four approaches simultaneously but to experiment with each; take note which ones generate the most traction for your recovery.

Positive Resource States

Trauma often rewires the brain, making it more sensitive to loss or uncertainty; however, this same ability to rewire can also be used to embrace moments of comfort, satisfaction, or pride. By engaging with these moments—or **positive resource states**—the brain receives new input for dopamine-driven learning. These instances begin to shift the brain from surviving to thriving.

The very circuits that wire fear and avoidance into your brain can be redirected toward growth with careful, consistent practice.

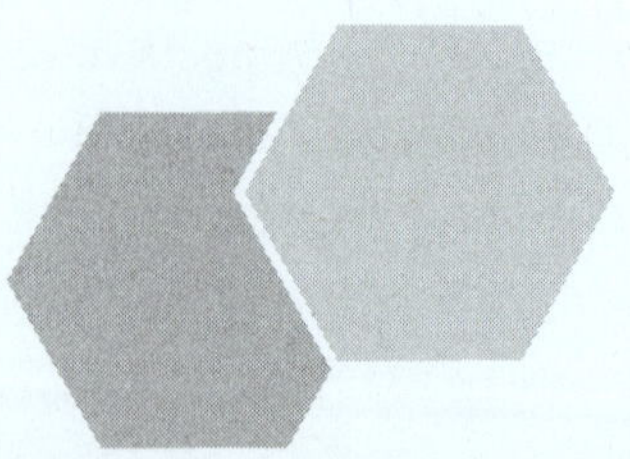

○ A CLOSER LOOK
Positive Resource State

A positive resource state is a mental, emotional, or physical experience of safety, calm, or well-being. It's the inner condition you move into: feeling grounded, held, or momentarily at ease.

For many, tapping into positive resource states may seem short-lived or hard to access at first. This isn't a personal failing; it reflects how your brain's neural pathways have been strengthened through repeated experience. When you've spent extended time in stress, anxiety, or disconnection, your nervous system becomes sensitized to threat and your dopamine circuits are primed to expect negative outcomes. Positive states feel unfamiliar because your brain hasn't yet learned to reliably access them—they haven't been reinforced frequently enough to become automatic. However, setting aside time to reflect on instances of support, confidence, or calm, even if rare, helps shift expectations. Paying attention to these moments gives dopamine the chance to reinforce them, nudging the mind toward openness and confidence.

Surprisingly, even the smallest sensations—a gentle breeze, a genuine compliment, the taste of a favorite food—can prompt healing. When you allow these moments to fill your awareness, no matter how seemingly insignificant they are, they gain importance in your mind. These subtle shifts, repeated over time, start to recalibrate emotional expectations and encourage resilience.

You can access these states whenever you want using **resourceful activities**. These are actions or practices you engage in to intentionally tap into a resource state, like taking a walk, calling a friend, or sitting in sunlight. The key is choosing activities that feel genuinely nourishing rather than obligatory—moments where you can notice a flicker of ease, pleasure, or connection. Even small acts, such as a mindful inhale, a gentle stretch, or a few kind words toward oneself can serve as radical signals to the nervous system that change is possible.

It helps when these moments arise naturally, tied to honest feelings and personal meaning. You may experience setbacks along the way, but each deliberate embrace of a positive resource state enforces your own resilience. Over time, as these activities accumulate, the dopamine system will update its expectations, making optimism and courage easier to access.

Therapeutic Support

Evidence-based therapies allow people to process trauma and recalibrate the nervous system in a structured way. Specific types of therapy like trauma-focused cognitive behavioral therapy, eye movement desensitization and reprocessing, and somatic experiencing have all been shown to be effective in helping survivors integrate difficult memories and become less reactive. These approaches work by creating conditions in which the brain can process threat signals in a safe space. As the brain relearns safety and gradually forms fresh associations with different experiences that were once upsetting, new possibilities open.

While this chapter primarily focuses on self-directed strategies, professional support can be a great aid in healing.

Intentional Discomfort and the Transformational Power of Courage

Real progress also involves gentle, supported steps toward discomfort. No matter how hard you may try to avoid them, challenging, uncomfortable moments are going to come up—especially when trauma has impacted your life. But studies show that braving small challenges—whether expressing vulnerability, revisiting difficult memories with support, or permitting a difficult feeling—puts dopamine to work for you. Every time this courage is met with understanding or affirmation, dopamine strengthens the link between effort and positive outcome, gradually reducing fear and boosting confidence.

It is this process of intentionally engaging with discomfort, rather than avoiding it, that creates the foundation for new beliefs about yourself and the world. Each time you weather a wave of discomfort and come out the other side, your brain receives critical evidence that you are not as fragile as the old patterns of trauma suggested.

In practice, engaging with discomfort might look like reaching out for help when you're used to handling everything alone, or revisiting a once-stressful environment and discovering you are able to cope. It can also look like trying something new. When a person attempts something different—like taking a new walking route or voicing a personal need—and this new step leads to a satisfying or empowering experience, dopamine pathways light up. This response then reinforces the idea that new possibilities can be safe and even gratifying. Through small wins and minor adjustments, you reinforce growth and courage, one dopamine boost at a time.

Keep in mind that embracing discomfort isn't about forcing progress or ignoring limits. As you take these steps, it's important to be kind to yourself and be flexible to changing circumstances. When minor setbacks occur (a missed goal, an awkward social encounter), the tendency might be to interpret them through the lens of old wounds—seeing them as proof that stepping out of your comfort zone is dangerous. However, deliberately redirecting your attention to the courage it took to make the effort creates a feedback loop where any attempt is recognized and valued regardless of the outcome. This compassionate way of viewing discomfort boosts dopamine and encourages future growth.

If you are healing from trauma, consider keeping a running log of even slight shifts in energy, mood, or confidence when embracing discomfort. By doing this, your brain can "see" the progress and keep those dopamine pathways attuned to curiosity and courage.

CASE STUDY
Pam and Artie—Rewiring after Loss

Pam and Artie found themselves drifting apart after the trauma of losing a child unexpectedly. Pam had been consumed with the loss. Decades later, it continued to affect every part of her life: persistent nightmares, intrusive memories, and a nervous system stuck on high alert. Any feelings of joy or hope seemed wrong. Meanwhile, Artie, although also suffering, was able to move forward in his goals and interests in the years since the loss.

Pam's brain, shaped by trauma, continuously replayed the accident and anticipated new dangers. Her dopamine system, once attuned to day-to-day pleasures and the unpredictable rewards of parenting, had become stuck in survival mode following her loss. Even small joys—a quiet coffee, time in the garden—were dulled by anxiety and guilt. Pam often spoke of life as being on "pause"; her routines felt empty of meaning. Artie sensed her drifting away and felt caught between worry and resignation. His efforts to comfort or "fix" her fears only seemed to widen the emotional gap between them.

We began by helping Pam find small sources of comfort that gave her a sense of control and hope. Our work centered on setting personalized routines for her—tending the flowers at her child's grave, gentle walks, listening to her favorite music—not to force optimism, but to give positive sensations time and space to grow. Pam learned that even brief moments of contentment could encourage her nervous system to adapt, bit by bit.

For Artie, the challenge was embracing companionship over problem-solving. He started joining Pam in quiet everyday moments—coffee outside, peaceful evenings, calm support on hard days—to help make togetherness feel secure again. Their connection deepened as they learned to navigate grief and celebrate small victories together. He started valuing honest exchange over quick solutions.

Over time, Pam experienced moments when anxiety eased and enjoyment returned. She still had setbacks, but she greeted them with newfound kindness and resolve. She noticed progress in small ways, like reaching for comfort instead of isolation. Each time she responded differently, Pam affirmed—through repetition and patience—that gradual healing was possible.

Artie slowly let go of his guard, discovering that listening actively and sharing experiences were far more healing than giving advice. As he and Pam rebuilt their relationship—one real conversation at a time—the distance faded. Together, they focused less on perfect outcomes and more on steady, authentic progress.

Pam and Artie's journey demonstrates that lasting strength comes not from avoiding hardship but from embracing new ways of connecting and coping. With the help of science, they built resilience and learned that, even when change feels daunting, the brain can adapt. Hope and healing are achievable.

Positive Prediction Error

The brain is constantly making predictions about what will happen next, then checking those predictions against reality. When reality doesn't match the prediction—whether better or worse—dopamine notes that discrepancy. This is called a prediction error, and it's one of the primary ways your nervous system learns and adapts. After trauma, the brain becomes skilled at *negative* prediction errors: expecting safety but encountering danger, anticipating connection but facing rejection. These experiences wire the brain to predict the worst as a form of protection.

For many who've faced trauma, a surprising moment of relief or kindness can stand out as a breakthrough. This is known as a positive prediction error: when reality turns out better than expected and rewires your brain as a result. Positive prediction errors reverse negative prediction errors.

A CLOSER LOOK
Positive Prediction Error

Your nervous system anticipated one outcome (usually negative or neutral based on past trauma), but something better actually happened. This mismatch between prediction and reality triggers a dopamine surge that marks the experience as significant and worth remembering.

Here's how it works in trauma recovery: After trauma, your brain becomes an expert at predicting the worst. Positive prediction error breaks that pattern by providing concrete evidence that your predictions were wrong. For example, say you expect a social gathering to feel overwhelming and isolating based on past trauma responses. Instead, you have a genuine conversation that leaves you feeling seen and less alone. Your brain registers: The prediction was wrong, and something good happened instead. That evidence then gets encoded by dopamine as a shift in what's possible for you. Each time reality spontaneously exceeds your low expectations, dopamine signals this as important learning, gradually shifting what your brain believes is possible.

It's often the simple moments of spontaneity—a gentle gesture, a brief respite from anxiety, or a reconnection with someone meaningful—that cause the greatest shift. These events help recalibrate the mind's view on what is possible, proving that relief and optimism can come even in ordinary circumstances.

Treating these unplanned moments as essential ingredients for healing gives the brain more opportunities to learn. Allowing for small shifts in routines, welcoming spontaneity, and pausing to acknowledge even minor new comforts help reinforce adaptive patterns. Over time, these moments accumulate, supporting openness, patience, and the willingness to try new experiences. For those overcoming trauma, positive prediction error becomes a key in reigniting energy and restoring a sense of possibility.

Practical Strategies: Building Your Healing Menu

Pause and practice: In Chapter 8, you created a Dopamine Menu designed to optimize motivation, focus, and everyday satisfaction. That menu assumed a stable mental baseline. However, trauma disrupts that baseline entirely. When the dopamine system has been reshaped by fear, hypervigilance, and avoidance, recovery requires a specialized healing menu. Here, you'll build a menu that prioritizes safety, nervous system regulation, and a gradual rewiring of the brain's threat responses. These strategies are tailored specifically to address the ways trauma dysregulates dopamine. You're not enhancing your nervous system; you're rebuilding it in the wake of trauma.

Once you've restored baseline stability and your dopamine system begins to respond more flexibly, you can integrate these healing practices with broader strategies from earlier chapters. But for now, focus on what your brain needs most: evidence of safety, moments of genuine ease, and the gradual rewiring of fear into possibility.

1. **Make micro-changes to spark big shifts.** Start with bite-sized actions—try a new snack, take a different route on a walk, or experiment with a creative hobby. Each novel experience can create a micro-boost of dopamine, reminding your brain that life is full of small surprises and pleasures.

2. **Savor the unexpected.** When something pleasant and unexpected happens (a smile from a passerby, a favorite song on the radio), pause and let yourself notice it fully. Give your brain time to register this positive prediction error.

3. **Resource rituals.** Establish daily or weekly routines that bring a sense of safety, self-compassion, and connection, whether through mindful movement, writing gratitude notes, or engaging in soothing sensory practices such as holding a warm cup of tea, taking a cool shower, applying lotion to your skin, or lighting a candle with a scent you find calming. Let these routines serve as reliable sources of comfort.

4. **Gently challenge avoidance.** Notice when you are pulling away from discomfort. Try approaching the feeling (rather than avoiding it) and pair that effort with support, encouragement, or a comforting resource.

5. **Strengthen connection.** Prioritize moments of genuine connection, no matter how brief. Laughter, warmth, or honest conversation can all stimulate dopamine and help rebuild trust in relationships and the world.

6. **Track progress.** Keep a record—a journal, a note on your phone, or a daily check-in—of small wins, unexpected joys, or moments of courage. Seeing your progress, even when incremental, reinforces the path toward healing.

Why this matters: Ultimately, the healing menu isn't static. It's an ever-evolving collection of strategies that can be expanded, adapted, and personalized as your resilience deepens and your needs shift.

CHAPTER SUMMARY

Trauma reshapes the brain's dopamine pathways, but the right strategies can help you tap into resilience and turn avoidance into engagement and healing. The following are the key points to remember from this chapter:

- Neuroplasticity allows the circuits that once reinforced fear and withdrawal responses to be gently rewired for hope, motivation, and authentic reward.
- Dopamine highlights new, positive experiences (especially unexpected ones), helping the brain update its predictions toward safety and possibility.
- Trauma survivors can often feel stuck because their dopamine system has learned to tune out pleasure and reinforce caution, not because of personal weakness.
- Small, surprising moments of joy and little wins provide powerful opportunities for the brain to notice progress and expand what feels possible.
- Consistent rituals, resource-building practices, and supportive relationships serve as daily catalysts for healing, prompting the release of dopamine as a reward for curiosity, agency, and meaningful connections.

PART 4

Into the New Frontier

Redefining Fulfillment, Enhancement, and Community

What if motivation, happiness, and human connection are even more adaptable than people believe? As this journey through the powers of dopamine reaches its final chapters, this part invites you to look into the future of dopamine. The world is changing quickly. Technologies that promise to boost focus and drive are being rapidly integrated into everyday life, sometimes before users can fully make sense of them. This raises not just practical but personal and ethical questions: Is it possible to use these new tools for growth without losing touch with what truly matters? What happens to connection and joy in a world shaped by brain science and technology?

Part 4 looks at both the possibilities and the risks of advancing technologies and dopamine manipulation. It is an invitation for you to explore the thrilling power of modern neuroscience—including topics like biohacking, nootropics, and self-directed improvement—while also being aware of the downsides and limits of these advances. Here, you will find stories of people who chased progress, only to discover that real fulfillment is deeper than any brain-boosting supplement or neural circuit. You'll also find clear, actionable guidance for navigating the ethical gray zones of modern technology and reconnecting with yourself and others along the way.

You have come this far because you are seeking meaningful change—not quick fixes. These chapters are about shaping a purposeful and connected life—not only as a unique individual, but as part of larger communities and societies. Whether you are curious about new brain research or looking to enjoy the simple things in life, the lesson remains the same: When growth is rooted in personal values and genuine connection, it is far more powerful (and sustainable) than technology alone.

Prepare to embrace a new definition of fulfillment: one that is based on courage, compassion, and community. The future holds extraordinary promise—not just for sharper focus or more neuroscience advancements, but for lasting satisfaction, happiness, and connection. This is where science meets purpose, and you become the creator of your most joyful life.

CHAPTER 15

Biohacking Dopamine

Nootropics, Neurotech, and the Ethics of Enhancement

As time goes on, scientific approaches to motivation, achievement, and emotional health will continue to go through big changes—not only through habits or personal resolve, but also by directly influencing the brain's chemistry. Dopamine, once a complete mystery, is now known as a driving force behind ambition, imagination, resilience, and daily engagement. Today's wave of biohacking, fueled by technological breakthroughs and a growing market of supplements and medications, promises far more than a quick boost. It reflects a major shift in how people think about cognitive enhancement, emotional regulation, and overall potential.

Across clinics, labs, and homes, noninvasive neurotechnology—including wearable stimulators, neurofeedback tools, and digital therapies—is being used alongside new medications to offer personalized ways for improving focus, emotional balance, and drive. But as innovation speeds up, so do the risks. Rapid gains in focus or energy can lead to tolerance, blunted emotions, disrupted sleep, or drug dependency. The ethics behind the use of such innovations is also changing, raising questions about authenticity, access, and the deeper costs of relying heavily on technology.

Neuroscience teaches that these new tools affect not only dopamine, but also the brain networks that shape attention, reward, and learning—meaning they can bring both profound benefits and unforeseen challenges. Real, lasting progress depends on thoughtful use, grounding innovation in basic habits such as good nutrition, movement, sleep, and social connection. As you read this chapter, you are invited to look at both the promise and limits of dopamine biohacking. Here, you will consider not only what is possible but also what is sustainable and meaningful as the world moves into a new wave of human enhancement.

Dopamine and Neurotechnology: Cutting-Edge Noninvasive Advances

Neuroscience is rapidly advancing, offering new opportunities to boost mental performance and emotional health without invasive surgeries or implants. Technologies that once seemed futuristic or limited to research labs are now being woven into the routines of everyday life. In the following sections, you will explore the different noninvasive advancements in brain science, from transcranial magnetic stimulation to AI neurofeedback.

Transcranial Magnetic Stimulation (TMS)

Transcranial magnetic stimulation (TMS) uses magnetic pulses to stimulate specific areas of the brain to increase the release of dopamine and improve focus and mood. TMS can deliver benefits within days, sometimes helping where medications have failed. More and more people are using this treatment to renew energy and creativity without the use of pharmaceuticals.

Transcranial Direct Current Stimulation (tDCS)

Transcranial direct current stimulation (tDCS) uses wearable devices to deliver gentle currents of electricity to specific circuits in the brain (typically those involved in attention, working memory, and learning). Lightweight sensors make tDCS accessible and user-friendly, allowing people in many fields to try this form of **cognitive training**. Those in education, research, clinical therapy, and even athletics and business are increasingly able to experiment with tDCS at home, in clinics, or in the workplace, improving attention and learning in these settings.

🔍 A CLOSER LOOK
Cognitive Training

Cognitive training refers to the structured mental exercises designed to strengthen skills like focus, memory, problem-solving, or processing speed. Cognitive training can include computerized brain games, attention drills, real-life strategy tasks, and more, all intended to drive neuroplasticity and optimize brain function.

Brain-Computer Interfaces and AI Neurofeedback

Wearable **electroencephalography** (**EEG**) devices and AI-driven **neurofeedback** platforms track brain wave patterns in real time, giving users immediate feedback about how activated, calm, or focused their brains are in that moment. By seeing these shifts as they happen, people can learn to adjust their breathing, posture, or attention and gradually train their brains toward more regulated, resilient states.

🔍 A CLOSER LOOK
EEG and Neurofeedback

EEG measures the brain's electrical activity through small sensors placed on the scalp. Different patterns of brain waves are associated with states such as deep focus, relaxed wakefulness, drowsiness, or stress.

Neurofeedback uses this information as a kind of "brain mirror." Software translates brain wave activity into simple visuals or sounds so users can practice shifting their state—rewarding patterns linked with calm focus and discouraging those linked with distraction or overload over time.

Minimally Invasive Innovations: DeepFocus and Adaptive Neurostimulation

Minimally invasive devices like DeepFocus use gentle scalp sensors and soft applicators that rest just inside the nostrils. Instead of anything being pushed far into the nose, these devices deliver safe, low-level electrical pulses to the nerves close to the nasal passages—a bit like a mild tingle. This targeted technique aims to activate specific parts of the brain connected to mood and motivation, offering an option that's comfortable, quick, and easy for people to use. Adaptive neurostimulation takes this further by using real-time feedback and AI algorithms to personalize the electrical pulses to each individual's brain patterns, adjusting intensity and frequency based on ongoing neural monitoring.

Many experts believe these technologies may help treat conditions ranging from depression to addiction, as clinicians can now guide treatment with greater accuracy. Wearable neurostimulators, smart apps, and personalized brain mapping are beginning to be used together—creating real-time support for dopamine pathways.

Nootropics: Enhancing Dopamine and Cognition

Nootropics are natural or synthetic substances that are taken to boost cognitive function, sharpen focus, and support healthy brain chemistry. These compounds can play a significant role in dopamine biohacking, helping to enhance attention, motivation, and mental energy in different ways. Both naturally derived supplements and pharmaceutical options may support dopamine signaling, but their effects can vary widely.

While some people use nootropics to sharpen their minds or boost productivity, it's important to understand their effects, safety, and interaction with your individual brain chemistry. Not every "brain booster" acts directly on dopamine, but many support pathways that shape attention, drive, or pleasure.

🔍 A CLOSER LOOK
Nootropics

Nootropics, sometimes called "smart drugs" or cognitive enhancers, include a range of compounds. Examples include caffeine (from coffee or tea), L-theanine (found in green tea), omega-3 fatty acids (from fish oil), prescription medicines for attention, and even cutting-edge lab-developed molecules (like racetams or newer compounds designed to optimize mitochondrial function and neuroprotection).

Common nootropics that can positively impact dopamine include:

- **L-tyrosine** and **N-acetyl L-tyrosine:** Aid dopamine production
- **L-theanine:** Promotes focused calm
- **Rhodiola:** May help make dopamine receptors more responsive to challenge and novelty
- **Citicoline:** Enhances neurotransmitter production and strengthens neural membranes, supporting cognitive energy and protection
- **Lion's mane mushroom:** Thought to restore dopamine balance and reinforce long-term brain adaptability; this nootropic is increasingly popular in biohacking circles
- **Caffeine:** Boosts alertness and drive, though its use should be balanced to avoid tolerance and overstimulation

Natural supplements also play a key role in biohacking, since they nourish the **gut microbiome** (the bacteria and other microorganisms living in the digestive tract that communicate directly with the brain and influence dopamine production, mood, and motivation).

🔍 A CLOSER LOOK
The Gut Microbiome

> The gut microbiome helps break down food, produce vitamins, and generate signaling molecules that talk to the brain through the gut–brain axis. Certain gut bacteria can influence how much dopamine and serotonin your body produces, as well as how sensitive your brain is to these chemicals. Diets rich in fiber, fermented foods, and diverse plant sources tend to support a healthier microbiome, which in turn can stabilize mood, energy, and motivation.

Key vitamins and minerals—B6, B9, D, magnesium, iron, zinc—act as nutrients to sustain dopamine activity and brain adaptability. **Omega-3s** (essential fatty acids crucial for brain health), particularly DHA and EPA, enhance the brain's structure and support flexible mood and stress responses.

🔍 A CLOSER LOOK
Omega-3s

> Omega-3s are a special type of healthy fat that the body can't make on its own—meaning you have to get them from foods like fish (salmon, sardines, tuna), walnuts, flaxseeds, and some plant oils. They help build brain cell membranes, reduce inflammation, and support healthy communication between brain cells. Getting enough omega-3s has been linked with better mood, sharper mental focus, and greater resilience to stress.

Modern supplement blends combine ingredients such as these to boost dopamine levels and provide many cognitive benefits. Whether the goal is to stay focused under pressure, balance emotions in stressful times, or recover after a period of high productivity, responsible use of supplements (and nootropics) can help you find mental clarity and build resilience in real time.

Enhancement Ethics: Navigating the Promise and Peril

Cognitive enhancement—whether through neurotechnology or nootropics—is transforming daily life, as well as brain health. With prescription aids, supplements, and noninvasive devices becoming more accessible than ever, more people can use them in learning, work, and creativity. Still, this accessibility also brings important ethical questions about how these tools shape choices, relationships, and society as a whole.

Fairness is at the core of the conversation. If powerful cognitive enhancers and neurotech tools are only available to some, will they create larger gaps between those who can afford such treatments and those who cannot—widening disparities in performance, opportunity, and access to reward? In education, athletics, and the workplace, questions about equity and privilege challenge existing systems, and it is just as important to consider how these technologies may either empower or exclude. An unequal distribution of these tools might reshape how talent and success are measured in society unless proactive steps are taken toward inclusivity.

Questions of authenticity are also emerging. When technology amplifies drive or creativity, what counts as genuine effort or true achievement? Some worry that persistent reliance on external aids could change how people relate to themselves and others. Will what matters most—inner growth, honest mastery, and real connection—be forgotten along the way?

In the wake of these questions, more are recognizing the need for transparency and clear guidelines. Stronger regulations and informed conversations across science, policy, and the public will be essential to protect safety, fairness, and trust as these tools evolve.

Ultimately, genuine health is about more than just maximizing output. As new tools promise more energy and focus, it's vital to pause for rest, reflection, and honest self-care. Science and compassion must go hand in hand—ensuring that enhancement truly supports resilient minds and thriving communities. As you explore new ways to boost your dopamine pathways in this book and beyond, keep this perspective in mind. Sustainable progress depends on remembering that well-being and purpose matter every bit as much as productivity.

Manipulating Motivation: The Possibilities and Downsides

These advanced biohacking tools offer sharper focus, renewed energy, and unexpected bursts of creativity—sometimes helping break through the very barriers that once held them back. Students preparing for exams, entrepreneurs launching bold ideas, and artists navigating creative ruts are discovering that the right enhancements can shift what feels possible.

Yet there's a trade-off. Chasing constant drive—primarily through repeated external boosts—can push the brain's reward system beyond its natural limits. Over time, it may take more stimulation or a higher supplement dose just to feel normal. Some find themselves running low on authentic motivation, feeling emotionally depleted, and wondering where their original spark has gone.

Recent studies also point to larger psychological costs of constantly tinkering with dopamine in these ways: disturbed sleep, increased anxiety, or a struggle to enjoy the simple pleasures that make life vibrant. Many discover that relentless improvement—without mindful limits— breeds burnout rather than fulfillment. True success means knowing when to use dopamine hacks and when to step back and honor cycles of rest, play, and meaningful relationships. Ultimately, real growth comes not from constant self-hacking, but from using these new tools wisely, with compassion for your own mind and life.

CASE STUDY
Yara—Pushing the Limits of Motivation

Yara, an entrepreneur, stepped into my office feeling both energized and uncertain. She had been combining supplements, smart devices, and innovative drugs in the hope of elevating her business and igniting her creativity. At first, she had enjoyed a rush of clarity and new ideas, feeling unstoppable. But soon, her sleep began to slip, and her emotions felt muted, the thrill replaced by worry and restlessness.

Driven to reclaim her sense of self, Yara shared her concerns: sleepless nights, constant agitation, and a growing disconnect from her deeper motivations. To move forward, we didn't just analyze her brain-hacking regimen—we mapped out her daily highs and lows, her stress triggers, and what truly mattered to her. Together, we unpacked the science, demystifying how dopamine cycles, neurostimulation, and supplements shaped her experiences.

Then, we crafted a flexible approach rooted in brain science, but also grounded in the rhythms of her everyday life. Yara practiced honoring sleep, carving out space for rest and play, and rediscovering small joys that felt genuine. We wove in gentle mindfulness, movement, and positive connection to retrain her reward circuits, creating a foundation for authentic, enduring resilience. As Yara rebuilt her routines, we talked often about what achievement meant to her—aligning ambition with authenticity and long-term well-being.

Looking back, Yara shared, "I thought biohacking would be like flipping a switch for constant productivity and high energy—the answer to every challenge. But I realized that just chasing dopamine wasn't enough. There was so much more to balance: sleep, connection, and my feelings about my achievements. The real breakthrough came when I gave myself space to recalibrate and found meaning in those quieter moments as well." Her openness during our sessions helped pave the way for deeper understanding and sustainable change.

Yara's story is a reminder that there's no shortcut to feeling truly empowered. Growth takes curiosity, patience, and a willingness to adapt. When science and self-reflection work in tandem, you acknowledge both your strengths *and* your limitations, creating space for bold achievement and genuine joy.

True success means knowing when to use dopamine hacks and when to step back and honor cycles of rest, play, and meaningful relationships.

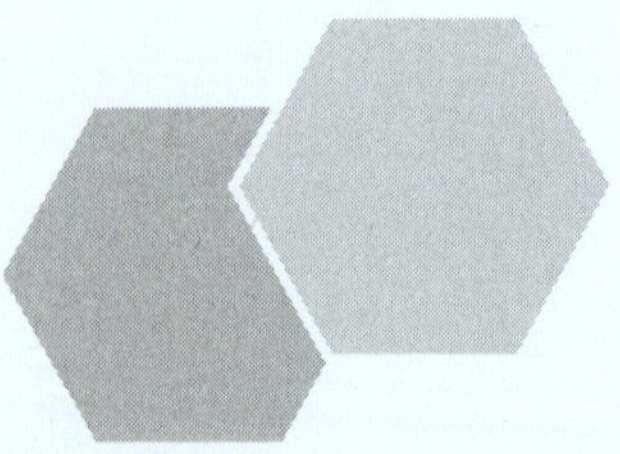

Biohacking in Practice: Integrating Techniques with Daily Life

Biohacking is most effective when it's integrated into daily routines rather than reserved for special situations. Real progress comes from a balanced approach—setting clear intentions and creating focused work windows, as well as carving out time for genuine rest and fun. Healthy boundaries foster both dopamine balance and steady emotional well-being, especially when life gets demanding.

Experimenting with different aids will help you blend biohacking into your everyday habits. Meanwhile, journaling or using tracking apps will allow you to note which habits are helpful and which ones need adjusting. Additionally, including movement, positive social exchanges, and gratitude rituals will reinforce the brain's natural reward circuits along the way.

Above all, sustainable success comes from integrating biohacking within the broader context of overall well-being. This means giving equal attention to rest, recovery, and reflection, honoring both the drive to grow and the need for balance. When technology and supplements work together with healthy habits and human connection, lasting motivation becomes not only attainable but also deeply rewarding. In the next section, you will learn how to integrate biohacking tools into your Dopamine Menu for daily growth.

Biohacks and the Dopamine Menu: Personalizing Your Path to Flourishing

Advances in neuroscience allow each individual to shape their Dopamine Menu into one that works with their unique rhythms and preferences. As you learned in Chapter 8, this isn't a one-size-fits-all recipe—it's a tool kit that builds motivation, creativity, focus, resilience, and even boundless joy. The best menu understands that your brain's needs are distinct and that neuroscience is as ever-changing as you are. Try routines that weave biohacks into scheduled movement, nutrition, and creative or social experiences. Stay curious and test which advancements help your brain light up in positive ways.

For added guidance as you are adapting your own menu, here are some examples of Dopamine Menus designed for different needs:

- **Creative Professional:** Early morning movement, protein-rich breakfast, bright workspace, scheduled social "idea jams," and afternoon technology breaks
- **Student:** Timed study blocks with music for focus, omega-3 supplement, sunlight exposure breaks, end-of-day gratitude journaling, and light L-theanine use before tests
- **Entrepreneur:** Mindful meditation intervals, walk-and-talk meetings, adaptive neurofeedback, rhodiola-based supplement stack, tech-free evenings
- **Busy Parent:** Family meal rituals, five-minute movement bursts, fun playlists for transitions, positivity journaling, and brief digital detox after bedtime
- **Retiree:** Tai chi or gentle yoga, creative hobby time, sensory-rich nature walks, small acts of kindness, connecting with friends weekly

Keep your menu adaptable. Sometimes, tech-forward neurofeedback may elevate productivity; at other times, journaling, movement, or spontaneous exploration may feel the most rewarding. Also remember to give yourself permission to revisit and revise your menu as life shifts—whether during periods of high stress, major transitions, or simple changes in health and priorities. At the heart of the Dopamine Menu is personalization and the understanding that progress isn't about perfection. You are invited to approach each biohack (and other dopamine-boosting practices) as an experiment, to reflect, and to celebrate every small win that signals growth or honest insight. Every habit chosen with intention, tested in a real-world context, and mindfully adjusted is a meaningful step toward living with greater engagement, harmony, and authenticity.

Looking Ahead: What the Next Decade of Advancement Holds for the Brain

In the decade ahead, advancements in brain science and technology will become more and more present in living rooms, classrooms, and workspaces. Smart tools powered by artificial intelligence and tailored to your unique brain will respond in real time to changing moods, energy, and goals. For example, AI-powered EEG headbands like Muse can detect shifts in your attention and stress levels in real time, then automatically prompt personalized interventions—from guided breathing exercises when cortisol spikes to dopamine-supporting activity recommendations based on your actual neural patterns. Wearable stimulators, carefully crafted supplements, and interactive digital therapies can work together to support dopamine in thoughtful ways.

You're entering a time of great choice—How much influence should neurotech have in shaping who you are, how you feel, and what you achieve? Alongside questions of fairness and safety, new concerns about privacy are emerging: Who owns your brain data, how is it stored, and what protections exist to keep it from being misused? You can expect more open and collaborative conversations about using these rapidly advancing technologies in families, schools, and workplaces.

Knowledge about brain health will broaden, welcoming advances such as microbiome support, improved sleep routines, and movement linked to digital tracking. The wisest paths will blend scientific breakthrough with age-old practices—good rest, nourishing relationships, and honest reflection. True empowerment isn't about doing more but about thriving with meaning and honoring what matters.

Practical Strategies: Ethical and Effective Enhancement

Pause and practice: Translating the science of dopamine and advancing neurotechnology into daily life means choosing practices that are both effective and responsible. Drawing on the dopamine basics from

Part 1 and Chapter 8, where you created your personalized Dopamine Menu, this chapter helps you thoughtfully adapt and refine those routines using the latest tools. Use this space to pause and try actionable strategies for working with your brain's natural strengths.

1. **Start with the essentials.** Establish foundational habits such as regular physical activity, consistent sleep routines, and balanced nutrition. Prioritize foods rich in protein, omega-3 fatty acids, and antioxidants to support dopamine production and protect brain health. As part of this foundation, also create structured daily rhythms with intentional periods of focused work, creative exploration, social connection, and restorative downtime. Use simple tools—calendars, timers, or mindful check-ins—to minimize distractions and stay focused.

2. **Integrate neurotech tools selectively.** Try wearable neurofeedback devices like Muse or EMOTIV, adaptive stimulation tools such as transcranial direct current stimulation (tDCS) devices like Halo, or guided meditation apps like Calm or Headspace for short periods. Track changes in your attention, mood, and energy to discover what works.

3. **Use supplements and nootropics thoughtfully.** Research the different ingredients and blends to discern what options may be most beneficial for you. Journal the effects, monitor tolerance or side effects, and consult professionals as needed to fine-tune these enhancements for your unique brain.

4. **Protect your well-being and values.** Include regular digital detoxes and relaxation practices—such as breath work, nature walks, or movement rituals—to maintain emotional balance and prevent overstimulation. Pause to ask whether an enhancement aligns with your personal goals and values, supports meaningful connection, or risks dependency or imbalance.

5. **Encourage ongoing learning.** Track results and adjust routines as life evolves. Share experiences and seek support by reaching out to

trusted peers or specialized coaches for feedback, inspiration, and troubleshooting. Growth thrives in community and conversation.

6. **Ask yourself these questions for safe dopamine enhancement.** What signs suggest I'm overusing supplements or neurotech? How can I avoid developing tolerance or emotional numbness? Are all nootropics and neurostimulation devices researched and safe for long-term use? What's the best way to track mood, energy, and sleep when experimenting? How do I strike a balance between motivation boosts and genuine rest and connection? Who should I consult before making significant changes to my enhancement regimen?

Why this matters: By integrating these practical strategies into daily life, you can harness dopamine biohacking and neurotech to create a vibrant, sustainable, and ethically grounded path to well-being and fulfillment.

CHAPTER SUMMARY

Grounded in the latest neuroscience and practical application, this chapter explored how biohacking dopamine can help you move beyond your limits—inviting curiosity, wise experimentation, and lasting growth. Here are the main takeaways from this chapter:

- Biohacking dopamine through neurotechnology and nootropics offers new ways to boost focus, energy, and creative potential.
- Both noninvasive and invasive enhancements should be used responsibly; do your research and consider individual needs, limits, and long-term effects.
- The future of brain health is highly personal—what works for one person may not work for another. Experimenting and adapting is essential.

- Authentic well-being requires striking a balance between biohacks and foundational habits, including sleep, movement, nutrition, and social connection.
- Ethical dilemmas—such as fairness, authenticity, and emotional health—are important things to consider, as enhancement tools reshape how you work, learn, and interact.

CHAPTER 16

The Dopamine Revolution

Igniting a Life of Lasting Fulfillment

The Dopamine Revolution is all about shifting from distraction and stress to a life filled with creativity, joy, and connection. It calls you to rethink what true fulfillment, community, and achievement mean for a healthier mind. This vision of wellness is not just a distant goal—it reflects an urgent need seen across cultures and backgrounds. In every space, people describe a sense of burnout, a lingering feeling of loneliness, and a relentless pursuit of expectations, while true happiness feels out of reach. Many describe a persistent sense that "something is missing"—an invisible "thing" separating outward success from genuine joy. As you've learned throughout this book, science proves that lasting motivation and contentment do not hinge on luck, random bursts of energy, or external validation. Instead, they come from deliberate decisions, purpose-aligned habits, ethical biohacking, and supportive environments that nurture your brain's urge for connection and ritual.

Drawing on the latest breakthroughs in neuroscience, as well as the wisdom of psychology, anthropology, and sociology, this chapter guides you through taking the Dopamine Revolution into your own hands. Grounded in real-world stories and actionable steps, you'll find dopamine-friendly routines that not only transform your own well-being but also ignite positive change within relationships, workplaces, and entire communities.

More than a movement, the Dopamine Revolution is an urgent invitation. It's time to redesign your daily life to honor what really matters: connection, purposeful achievement, and the regular practice of happiness. By embracing bold, restorative changes and reclaiming what has been lost, you will take huge steps toward a future that is resilient and thriving.

The Future of Connection: The Dopamine Revolution and Brain Health

For generations, connection was regarded as an enjoyable backdrop to individual pursuits—a luxury rather than a necessity. Yet studies in neuroscience are continuing to reveal that genuine bonds are essential not only for cognitive health, but also for emotional stability and even immune protection. Within the Dopamine Revolution, enduring relationships stand as a primary pillar of holistic wellness. These connections shape the most robust and adaptive brains.

Unfortunately, as you learned in Part 2, modern life regularly pulls people away from meaningful connection. Social media, engineered addiction (algorithms and design features deliberately crafted to maximize engagement and psychological dependence), and the pursuit for more have impacted many relationships, leading to anxiety and loneliness.

Building genuine connection as you move into the future of dopamine science will require a redesign of your everyday routines to foster meaningful presence and social engagement. This can start with simple, intentional rituals, which can include:

- A heartfelt morning greeting
- A **gratitude circle** at work
- Device-free family meals
- Regular time devoted to laughter and reflection

A CLOSER LOOK
Gratitude Circle

A gratitude circle is a simple practice where a group briefly shares something they appreciate: a person, a small win, or a moment of meaning. This can happen at the start of a team meeting, around the dinner table, or in a weekly check-in with friends. The goal is not forced positivity, but helping the brain notice safety, support, and progress. Similar connection rituals include "Rose, Thorn, and Bud" check-ins, where everyone shares one positive, one challenge, and one thing they are looking forward to.

You can also boost the sense of mutual support and shared meaning that your brain craves through things like:

- Community events
- Mentorship
- Volunteering
- Group activities such as walking clubs or collective dance

To reimagine connection is not to gain more acquaintances or online followers. It is to cultivate real, reciprocal relationships across every sphere—family, workplace, and neighborhood.

The Future of Productivity: The Dopamine Revolution and a Healthier Mind

In recent years, productivity has been presented as constant output and digital connectivity, along with external recognition. However, as explored in Part 3, discoveries in neuroscience and organizational psychology reveal that this fast-paced approach and emphasis on more can harm brain health, reduce long-term motivation, and erode genuine fulfillment. The real power of productivity is not found in constant activity, but in intentional effort, mindful renewal, and creative engagement. Dopamine flourishes when you allow space for your personal values, natural cycles of inspiration and rest, and creativity.

To reimagine productivity, you must shift your focus from "doing more" to "doing what matters." This involves setting clear intentions, minimizing distractions, and engaging in purposeful work. Design daily and weekly rituals that include all of the following:

- Focused sprints
- Regular renewal
- Authentic feedback

You can also use various methods that consistently activate your reward pathways. These include:

- **Time-blocking**, where you divide your day into set periods or "blocks" dedicated to specific tasks or types of work
- The "deep work" you learned about in Chapter 5
- The **Pomodoro Technique**, which structures tasks into short, focused segments, traditionally twenty-five minutes of deep work followed by a five-minute break, repeated in cycles with a longer break after several rounds

A CLOSER LOOK
The Pomodoro Technique

The brain is not built for uninterrupted, hours-long hyperfocus. Short work sprints with planned breaks help regulate dopamine and stress hormones. During each focused block, your brain gets a clear goal and a defined end point, which boosts task-related dopamine and makes it easier to start. The brief rest periods then give your nervous system time to reset, preventing the crash that comes from grinding nonstop. Over time, this rhythm can improve attention, reduce procrastination, and make productive effort feel more sustainable and less punishing.

Within the Dopamine Revolution, real productivity nurtures healthy minds, joyful effort, and enduring success. By combining the drive for achievement with intentional times of rest, you will unlock sharper focus, deeper satisfaction, and a creative resilience that shines in every part of your life.

The Future of Happiness: The Dopamine Revolution and True Fulfillment

Sustained happiness—the hallmark of the Dopamine Revolution—is not about accumulating or accomplishing more; it is about being deeply present and engaged in life. As studies continue to show, engaging in activities you find personally meaningful stimulates brain regions associated with well-being, sense of meaning, and engagement.

Just look to joyful, long-lived cultures for proof: Okinawans in Japan who practice *ikigai*; celebratory gatherings in the Mediterranean villages of Italy, Greece, and Spain; or those who embrace the ritual of *hygge* in Denmark. Each of these traditions weaves meaning, connection, and simple pleasure into daily life, giving the brain steady, nourishing dopamine rather than constant spikes and crashes.

A CLOSER LOOK
Okinawan *Ikigai* and *Hygge*

Okinawan *ikigai* is a daily ritual of doing what has purpose—caring for family, nurturing a garden, or practicing a craft.

Hygge, a term referring to a feeling of coziness, turns ordinary moments—warm lighting, shared meals, relaxed conversations—into reliable comfort and belonging.

These happy communities reveal key lessons: Joy flourishes where gratitude, storytelling, and play are woven into daily life. These traditions activate the brain's reward and emotion regulation centers, fostering both individual satisfaction as well as collective achievement.

Other activities that will nurture your dopamine pathways and foster happiness include:

- Helping others
- Mastering new skills
- Facing shared challenges

Much more than pleasurable, these practices bring lasting well-being. When fulfillment becomes your compass, every shared moment of connection, growth, and contribution strengthens your brain for resilience and builds a more joyful, purposeful life.

Why Now: The Costs of Staying in the Past

Despite living in an age of advancing technologies and expanding opportunities, society is facing a growing, but often overlooked, issue. As you uncovered in Part 2, loneliness, stress-related illness, and burnout are becoming more and more common. Surveys consistently report that nearly half of adults experience feelings of isolation each week, while workplaces struggle with chronic fatigue, disengagement, and diminishing morale. Although digital networks were designed to bring people closer, for many, these tools have only amplified the things that alienate and isolate: comparison and distraction. Psychologists and neuroscientists warn that chronic disconnection will increase your vulnerability to both mental and physical health issues over time, including early cognitive decline and mortality.

At this pivotal crossroads between ambition and genuine fulfilment, the Dopamine Revolution is an invitation to reengineer how you connect, collaborate, and recover—laying the foundation for a more fulfilling future.

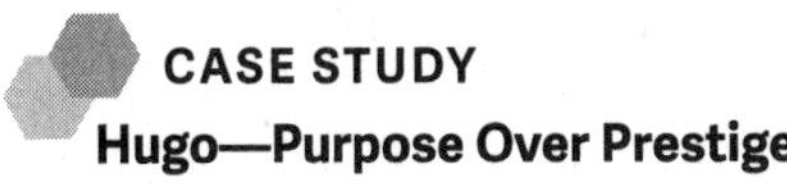

CASE STUDY
Hugo—Purpose Over Prestige

Hugo came to me at a crossroad. He was a highly accomplished college professor with a distinguished career, yet he quietly struggled with exhaustion, isolation, and a persistent sense that something essential was missing. For years, accolades and expertise had brought recognition, but not the deeper satisfaction he craved. "How can a life so full feel this empty?" he asked during our first conversation, echoing a feeling I had heard from countless clients before him.

Together, we explored the neuroscience behind motivation and belonging. Hugo realized that his routines, built around endless deadlines, solitary research, and the pursuit of academic prestige, were starving his brain's reward circuits and dampening his creativity. Social isolation had become the norm, yet the impact on his mood, energy, and drive was profound. Our sessions focused on how deliberately fostering connection—rather than chasing achievement—could transform not just his professional life, but his sense of fulfillment.

Instead of ignoring these insights, Hugo made bold changes. He introduced gratitude circles before faculty meetings, scheduled informal walks with colleagues, and found ways to invite authentic conversations—not only about research, but also about life, dreams, and setbacks. He created more open spaces for collaboration, reflection, and collective celebration on campus. He also made small but equally impactful shifts: listening deeply, acknowledging wins, sharing stories, and simply slowing down to savor real moments together.

The results were transformative. Hugo found renewed energy, stronger relationships, and a sense of belonging that reignited his love for teaching and learning. His colleagues also reported lowered stress, along with increased creativity and teamwork. Absenteeism dropped, and meetings became lively exchanges rather than obligatory tasks. "I'm finally living what I've dedicated my life to studying," Hugo reflected. "Fulfillment isn't the product of working harder—it's what happens when we build real connection and purpose into every day."

Hugo's story is more than an individual breakthrough—it's a testament to the Dopamine Revolution. His changes inspire others to ask: "Where am I substituting busyness for belonging, and what happens when I choose connection instead?"

Practical Strategies: Building Dopamine-Friendly Habits and Communities

Pause and practice: Embracing the Dopamine Revolution begins with simple, intentional action—reshaping daily life so that connection, meaning, and joy become living routines. Take time now to use the following strategies for fostering dopamine-friendly habits at home, work, and in community life:

1. **Start small.** Begin by introducing little rituals that spark engagement and connection. Share a gratitude moment before breakfast, take a movement break with colleagues, or schedule a brief weekly call just to celebrate small wins. Neuroscience and anthropology show that even brief rituals—such as greetings, shared laughter, and communal meals—lay a durable foundation for trust and belonging.

2. **Design your spaces for human connection.** Rearrange rooms and work areas to encourage spontaneous conversation and creative teamwork. At home, gather for evening reflections or family games. In meetings, invite everyone to share something before getting to formal agenda items. Schools and offices can offer creative lounges, green spaces, or shared cafés where ideas and relationships flourish organically.

3. **Cultivate radical inclusivity.** Real fulfillment demands inclusivity. Invite newcomers into group traditions and rotate leadership for rituals, ensuring that all perspectives are represented and every member feels valued. In neighborhoods, consider starting potlucks, walking clubs, or volunteer projects that foster diverse friendships and generate collective wisdom.

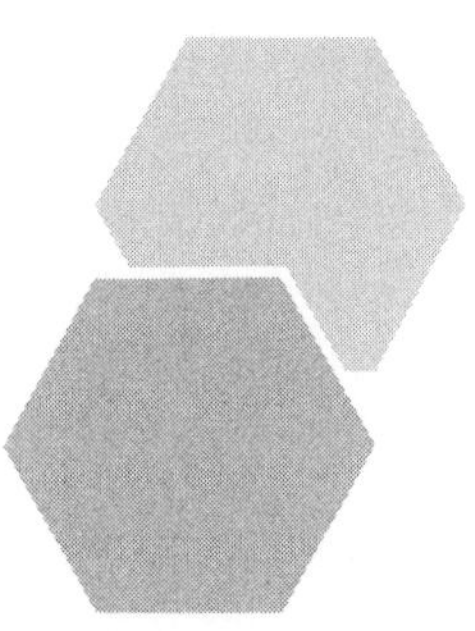

When fulfillment becomes your compass, every shared moment of connection, growth, and contribution strengthens your brain for resilience and builds a more joyful, purposeful life.

4. **Make movement and play essential.** Evolution shaped your brain for active, dynamic days. Honor this ancient need by building movement into your regular routines: Try stretching, dancing, taking outdoor strolls, or even engaging in playful brainstorming sessions at work. Schedule "tech-free" hours at home, embracing creative play (such as imaginative storytelling, cooking challenge nights, card games, board games, or backyard obstacle courses) to let your dopamine circuits reset and boost imagination.

5. **Celebrate and reflect regularly.** Anchor your dopamine-boosting routines with regular celebration and reflection. Host monthly storytelling evenings, gratitude circles, or "open mic" sessions where you get to honor achievements and lessons learned. These shared rituals prevent stress and spark lasting optimism by activating reward pathways and deepening social bonds.

6. **Lead by example.** Ideal change starts with leaders who prioritize kindness and collaboration. Be transparent about challenges, celebrate others' growth, and set healthy boundaries—showing that real strength is rooted in empathy, vulnerability, and mutual support. At every level, role models inspire a culture of trust.

7. **Troubleshoot barriers with compassion and creativity.** Expect to feel resistance or awkwardness as your old habits give way to new ones. Practice gentle persistence: Ask for feedback, make space for hesitation, and keep celebrations low-pressure. Belonging grows through repeated, genuine acts—not grand gestures.

Why this matters: You know you have embraced the Dopamine Revolution when daily life is intentionally designed around inclusion, laughter, purpose, and creative collaboration. The ripple effect is profound: Your brain becomes more resilient, your relationships deepen, and your sense of fulfillment shifts from wishful thinking to reality.

CHAPTER SUMMARY

As you navigate this pivotal time in human development, it becomes clear that the path to a more fulfilling life is rooted in how you nurture your brain and strengthen your connections to others. The promise of the Dopamine Revolution is not just personal transformation, but a ripple effect that uplifts families, organizations, and entire communities. As you go forward in your journey with dopamine, keep these takeaways from this chapter in mind:

- Authentic connection is foundational to good brain health, motivation, emotional resilience, and creativity.
- Actual productivity supports both achievement *and* renewal, blending focused effort with restorative rituals and meaningful collaboration.
- Lasting happiness comes from engagement, gratitude, growth, and shared purpose—not quick dopamine spikes.
- Movement and play re-energize minds, reduce stress, and unlock new levels of creativity and problem-solving.
- Leading by example—modeling kindness, vulnerability, and celebration—accelerates positive change in any environment.
- The Dopamine Revolution is open to everyone; embrace it.

Your Dopamine Blueprint for Lasting Change

You've reached the end of this book, but the truth is, you're at the beginning of something significant. You now have knowledge that many people spend their entire lives searching for without ever finding: the actual science behind why you do what you do, want what you want, and struggle where you struggle.

Throughout these chapters, you've journeyed deep into the intricate world of dopamine and discovered that this chemical messenger shapes nearly every aspect of your existence. You've learned what dopamine truly is and how it orchestrates your motivation, desire, and reward systems. You've dismantled the myths that keep people trapped in confusion, understanding now that dopamine isn't just about pleasure seeking but about something far more nuanced. It is the delicate balance between satisfaction and craving, between fulfillment and emptiness.

You've met real people in these pages. Johan, who couldn't understand why professional success left him feeling hollow until he discovered his dopamine system was dysregulated by constantly chasing achievement. Anna, who found herself feeling increasingly lonely despite constantly connecting to social media—until she started fostering relationships beyond the screen. Mei, whose anxiety lifted once she understood how her surroundings discouraged her personal goals rather than invited them. These weren't exceptional people with unusual willpower or special advantages. They were struggling, just like you might be. They felt stuck in patterns they couldn't break, reaching for quick fixes that offered momentary relief, only to be followed by deeper dissatisfaction. What transformed their lives was exactly what you now hold: understanding paired with action.

Let me be direct. Those old neural pathways you've carved over years or even decades? They're still there. Your brain doesn't delete them. That deeply embedded wiring that learned to seek dopamine from scrolling social media, having unhealthy relationships, overworking yourself, or whatever your particular pattern happens to be—those pathways remain ready to reactivate. And sometimes they will reactivate. You'll find yourself pulled back toward old behaviors, feeling that familiar dopamine surge, only to crash afterward into regret, emptiness, or shame.

This will happen. Probably more than once.

But here's what's fundamentally different now: You understand the mechanism. When you feel that pull toward a quick dopamine hit, you'll recognize it as your brain running an outdated program. When you do slip back into old patterns, you won't have to spiral into that same dark place where you once felt utterly lost and powerless. You can pause, remember what you've learned (or even revisit certain chapters whenever you might need a refresher), and redirect yourself using the insights and strategies provided throughout this book. Your brain has already begun rewiring itself simply by understanding these concepts. New neural pathways have formed. The knowledge sits ready in your memory, accessible whenever you need it.

You can now recognize your triggers before they hijack you. You can identify when your dopamine baseline needs support versus when you're genuinely thriving. You have practical, science-backed approaches to cultivate sustainable motivation and build a life that delivers lasting fulfillment rather than fleeting pleasure followed by emptiness.

This is your dopamine code. Your personal road map to a genuinely happy, productive, and meaningful life. Take what you've learned and implement it through small, consistent choices. Practice self-compassion when you stumble. Celebrate your progress. Remember that transformation comes not from perfection but from persistent, conscious effort.

You've always had the capacity for change within you. Now you know how to unlock it.

Additional Resources

To keep exploring how dopamine shapes motivation, mood, and everyday joy, the following resources offer more information. These books, websites, and podcasts provide practical guidance, deeper science, and real-world tools to help you keep refining your own Dopamine Menu and building a happier, more sustainable life.

Books

Dopamine Nation: Finding Balance in the Age of Indulgence by Anna Lembke

The Molecule of More by Daniel Z. Lieberman and Michael E. Long

Habits of a Happy Brain by Loretta Graziano Breuning

Websites

Dopamine: The Pathway to Pleasure by Harvard Health
www.health.harvard.edu/mind-and-mood/dopamine-the-pathway-to-pleasure

What Is Dopamine? by Mental Health America
https://mhanational.org/resources/what-is-dopamine

Dopamine: What It Is & What It Does by WebMD
www.webmd.com/mental-health/what-is-dopamine

Apps

Roots (phone habits)

Headspace (mental health)

Insight Timer (meditation)

Podcasts

Huberman Lab with Andrew Huberman

On Purpose with Jay Shetty

The mindbodygreen podcast with Jason Wachob

Index

About the Author

Dr. Sydney Ceruto, founder and CEO of MindLAB Neuroscience, has been at the forefront of integrating neuroscience into personal and professional coaching for more than two decades. With two master's degrees in psychology and two PhDs in behavioral and cognitive neuroscience, Dr. Ceruto is widely recognized as an expert in the field of neuroscience. She practices personal coaching across the globe, with offices in New York, Miami, Beverly Hills, and Lisbon, Portugal. Dr. Ceruto is also a *Forbes* Coaches Council executive contributor, where she shares insights on neuroscience and leadership, and she has earned numerous prestigious awards and recognitions, including a 2024 Lifetime Achievement Award from the World Coaching Congress and a 2022 induction into the International Society of Female Professionals as one of the World's Top 3 Best Life Coaches.